THE BEST C

TOP SHOPS®

PLANNING, OUTFITTING, AND UTILIZING YOUR FARM SHOP

MW01612437

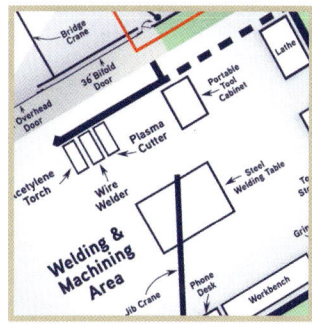
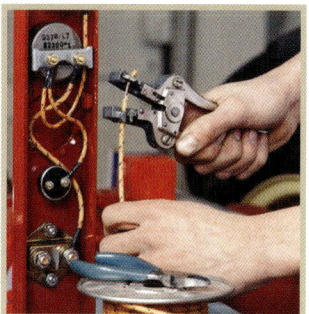

Successful Farming Magazine
Publisher: Scott Mortimer
Editor in Chief: Loren Kruse
New Product Manager: Diana Willits
Consumer Marketing: Brenda Torsky

The Best Of Successful Farming Top Shops
Editor: Paula Barbour
Art Director: Mark McManus
Editorial Director: Dave Mowitz
Production Manager: Pamela Garbett

Senior Vice President: Doug Olson

Vice President: Tom Davis

The Best Of Successful Farming Top Shops (ISBN 978-0-696-24303-5) is published by *Successful Farming* Magazine—Meredith Corporation, 1716 Locust Street, Des Moines, IA 50309-3023. Copyright Meredith Corporation 2009. All rights reserved. Write to *Top Shops* at 1716 Locust Street, LS 253, Des Moines, IA 50309-3023, or e-mail us at dave.mowitz@meredith.com.

ALL ABOUT
TOP SHOPS®

I have got to have the best job at *Successful Farming* magazine. True, the entire staff at the magazine enjoys serving the best people in the world in farmers. But my job duties provide the opportunity to feature the most innovative shops and shop inventions in the country, all of which are featured in the book you are now holding!

Farmer ingenuity comes to full blossom in shops as you will discover in this book. I am constantly amazed by the innovations I find when visiting farm shops. Besides serving as the headquarters of an operation and natural meeting place for salespeople and friends alike, shops are the structures where many farmers' dreams come true either in the form of a well designed floor plan, a first-rate homebuilt tool, or comfortable work space.

CREATED FROM NECESSITY, BUILT OUT OF DREAMS

Many of those dreams begin on the cold dirt floor of a makeshift shed-turned-shop during a frozen bearing replacement, a busted shaft repair, or as a broken weld is fixed. Frozen fingers and a sore back leave you yearning for heat, a hoist, wide workbenches, and maybe an office to relax in.

Certainly shops have become the place where many farmers' concepts come to life in the form of a well-designed and well-equipped work environment. Repair bills shrink when they are encouraged to do more of their own repair and maintenance. And equipment that is well maintained is far less apt to break down in the middle of a season.

Shops have also become miniature manufacturing facilities. That sprayer a farmer has yearned to own comes to life on a shop floor. And many farm shops service a thriving sideline business by providing repairs, building machinery, or selling supplies.

THE VALUE OF A GOOD SHOP IS INCALCULABLE

Putting a value on a good shop is next to impossible – this I have discovered in years of writing about such structures. Oftentimes a shop may not appear to be economical on the books. And it's tough to convince a skeptical banker that a new shop will put more dollars to your bottom line. But what's the cost of days lost during planting to a breakdown that could have been fixed in an hour in a good shop? How many dollars could be saved doing repair and maintenance chores at home rather than paying for the work at a machinery dealer's shop? What's the real price of frozen fingers and a sore back?

Dave Mowitz is Machinery Director at *Successful Farming* Magazine and host of the "Machinery Show" on RFD-TV.

Greg "Machinery Pete" Peterson (center) discusses how critical proper preventive maintenance and regular cleaning are for improving used machinery's resale value.

INVALUABLE GUIDE TO SHOP DESIGN, USE

This book offers excellent advice on shop design and illustrates facilities created by farmers who know the value of a good shop. And we've also included a great many examples of farmer-built shop tools, which were not only built for less cost than their store-bought alternatives but were also engineered to last longer while providing a wide variety of features.

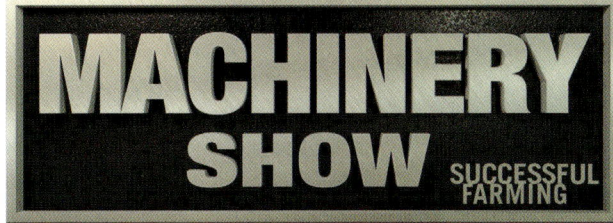

You are also welcome to join me on the "Successful Farming Machinery Show" – a half-hour program that airs several times each week on RFD-TV (check your television directory for program times and stations). This, the most popular television program in agriculture in the country, features outstanding shops and shop inventions, many of which have won top honors in *Successful Farming* magazine's Top Shops® Contest.

Dave Mowitz
Machinery Director
Successful Farming Magazine
Host, "Machinery Show"

A "Machinery Show" film crew captures the features of an award-winning Top Shops® service truck in Tulare, California.

Shop sage Roger Welsch imparts his wisdom on shop tool use – and misuse – such as the inappropriate application of a sandblaster, which proved capable of not only removing grime and paint (seen in the picture at left) but also better parts of a carburetor (right).

THE BEST OF SUCCESSFUL FARMING.

TOP SHOPS.

PLANNING, OUTFITTING, AND UTILIZING YOUR FARM SHOP

TABLE OF CONTENTS

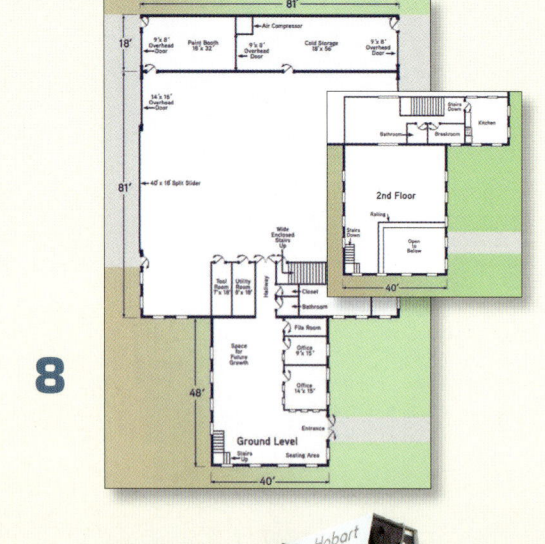

8

30

44

50

86

90

98

115

127

A Second Life as a Shop

Barn conversion creates a shop at half the cost of a new building

By Dave Mowitz

Nancy and Randy Lackender still use the side extension of their barn for hay storage. The structure's main room (behind the hay trailer) was converted into a shop.

Several years ago, Randy and Nancy Lackender had to make a tough decision that is all too common for farmers today: What would they do with their farm's barn?

The majestic structure was constructed by Randy's grandfather in 1915 on the family's farm near Iowa City, Iowa. Originally used as a dairy and cattle barn, this generation of Lackenders had utilized it for farrowing and finishing hogs.

WHAT TO DO WITH AN IDLE BARN

But the Lackenders' fabrication business continued to grow. Randy builds and sells a line of skid steer loader backhoes and buckets. For more information on Lackender Fabrication, go to www.lackender-fab.com.

"When our skid steer backhoe business grew, we had less time to spend on livestock. So we stopped raising hogs altogether, and the barn sat idle," Randy recalls. "We were fortunate because my dad had put on a new roof and covered the barn with metal. He'd always kept the barn well maintained. As such, structurally, the building was in great shape."

The Lackenders had noticed articles in *Successful Farming* magazine featuring converted barns. Several of those stories illustrated how farmers replaced low-level hay mows with elevated ceilings that would accommodate modern machinery. "That hay mow is the major hurdle to overcome when renovating a barn such as ours for a shop," Randy adds.

Seeing similar barn conversions encouraged the Lackenders to undertake the conversion project in 2002.

The work began with hanging prefabricated roofing trusses under the barn's existing roof. The old hay mow was kept intact to act as scaffolding to making installing the new trusses an easier, safer chore.

SEALED OFF WITH INSULATION

The purpose for the trusses was to replace the mow floor, act as the shop's future ceiling, and shore up the structure. Made to span the width of the barn's main room, the trusses rest on and are attached to the barn's roof sill plate. "After the trusses were installed, we removed the mow floor," Randy explains.

Metal roofing was then attached to the bottom of the trusses to create the ceiling. "After that I had approximately 1 foot of loose insulation blown across the top of this ceiling," Randy recalls. "And the shop's sidewalls were insulated utilizing sprayed-on urethane foam to an approximate depth of 1 to 2 inches. This provided an approximate insulating value of R-15 to R-20 in the walls."

Utilizing the sprayed-on insulation allowed Lackender to keep the barn's original wall beam exposed. Another advantage in using this type of insulation on the sidewalls is that the substance completely seals off the walls, Randy explains. Plus, you can paint the foam after it has hardened.

To finish off the shop, Randy covered the bottom half of the

Photographs: Ron Van Zee

sidewalls with planks salvaged from the old mow floor. Those planks, the urethane foam exposed on the upper walls, and the ceiling are painted white to make the area brighter.

BASE FOR A FUTURE HOIST

Previous to painting, the Lackenders had a 5-inch-thick concrete floor poured in the future shop. Before that floor was poured, Randy dug a hole 7 feet deep and approximately 5×5 feet in size. This hole would be the foundation for a future swinging hoist.

Next, he positioned an 8-inch-diameter pipe (with a ⅜-inch sidewall thickness) upright and in the center of that hole to serve as a support standard for a future swinging jib crane. The hole was then filled with concrete "providing a super-solid base for the hoist," Randy explains. "That foundation also allowed me to not have to attach the hoist's vertical support pipe to the structure's ceiling."

CRANE SERVICES THE SHOP

Randy fabricated the crane using 10×5-inch I-beams. The arm that extends out 16 feet into the shop is secured to its vertical base using ¼-inch plate steel gussets. Structural gussets were made from ¼-inch steel plate.

The hoist's base rides on homemade hinges that employ 1½-inch steel pins. "I didn't use bearings on the hinges, nor are they greasable," Randy says. "The hoist swings just as easily without them."

The crane was then topped off with a 1-ton hoist that swings in an arc, which covers a huge swath of the shop's floor and also reaches outdoors.

One of the advantages to renovating the barn is that all parts, lubricants, and miscellaneous storage can be tucked away in adjoining rooms.

"For example, the old feed room at the west end of the shop acts as parts storage," Randy says. "Double doors on the side of the shop lead to the old hog nursery, which provides ready access for additional storage. Using these adjoining rooms certainly helps to eliminate clutter in the shop."

DOOR CHOICE IS IMPORTANT

The main access to the shop is through a 15×20-foot sliding door located on a side wall.

"We selected a sliding door, as opposed to an overhead-operating unit, to help preserve the look of the barn," Randy explains.

"I have another shop that accommodates the fabrication business that has larger doors. So that means I can take my combine to that building to be worked on. However, this door has worked to accommodate all the farm's machinery."

FLOOR HOIST INVALUABLE

One of the best investments the Lackenders made in creating the shop was buying and installing a floor hoist. "We found this in a service station that was being torn down. The hydraulic rams are operated using shop air," Randy explains. "This is the best thing I put in the shop. I can't believe how much I use the thing for just simple service and maintenance."

Randy estimates the entire cost of renovating the barn totaled approximately half the expense of erecting a new building. "Plus, I preserved an invaluable piece of our family's heritage," he adds with pride. ∎

After finishing out the ceiling and walls, Randy Lackender painted the entire shop white to better reflect light throughout the room.

Randy utilized the old hay mow as scaffolding when installing the barn's new ceiling trusses. After the trusses were secured, the floor was taken out.

Prefabricated roof trusses built to fit the width of the barn were attached to the roof sill plate. Then the ceiling was finished with 1 foot of blown-in insulation.

Shop Design Winner

One complex is home to all machinery maintenance and office chores

By Dave Mowitz, Machinery Director

The Schaffers' two-story structure houses all their offices (on the left) and a massive shop (on the right).

When tragedy struck and flames consumed a dairy free-stall barn Patrick and Sarah Schaffer were remodeling into a much-needed shop, the couple took a long look at their needs. Armed with insurance money from the fire, the willingness to do much of the finished construction work themselves, and a barn-full of great ideas they garnered from other farm shops, the Eau Claire, Wisconsin, operators created a comprehensive complex. This structure meets their needs from repair and maintenance chores to running the business of a growing farm operation.

DETAILED PLANNING

What is so striking about the Schaffer farm complex is how complete their planning was prior to the erection of the building. Examine the placement of rooms throughout the complex (see schematic on page 9). Note how the office area is housed under its own roof in a building that T's into the structure housing the shop. This arrangement allows the business offices to be kept separate from the shop. In doing so, it excludes shop noise and dirt. Yet the office is quickly accessible from the shop through an internal door.

Also note the cold storage area for parts, supplies, and a paint booth in the addition of the 18-foot-wide lean-to structure attached to the side of the shop building's 88-foot-long wall. This creates extra storage space that keeps their shop floor clean at a relatively cheap cost.

Most of all, note the massive 63-foot-wide unobstructed work space in the shop. Patrick refers to this area as "a full-access design. I wanted to create a large area to accommodate a wide variety of equipment at once," he explains. "The use of multiple doors on both ends of the shop floor lets a combine or semi truck move in or out without having to remove other equipment on the floor. By making our shop floor so easily accessible, we can work on several projects at once. This has increased our efficiency since we don't have to move our projects outdoors or to storage if we're waiting on parts, for example."

Key to this unobstructed work floor area was the creation of multiple storage rooms. All parts, supplies, and tools are kept in these rooms – away from the shop floor.

Between the utility room and parts storage room is the hallway that connects the shop to the office. Off of this hallway is the office's bathroom and closet.

Patrick located a separate office in the shop for access by employees. All shop and repair manuals, a laptop computer, and work boards are located here. Extra shop storage is found in the second floor loft along with an employee bathroom, breakroom, and kitchen.

The addition of an insulated lean-to structure on the side of the shop makes room for cold storage of

Patrick and Sarah Schaffer's family includes daughters (from left) Alexis, 11, Samantha, 5, and Natalie, 13 (standing).

lubricants as well as temporary parts racks, a paint booth, and work space separate from the shop.

Finally, the Schaffer complex provides ample room for business offices. "We left the entrance to the office open to the ceiling to create a spacious area," Patrick points out. "The second floor in the office structure is now used for storage. But this provides space either for future offices or for a large meeting room." ■

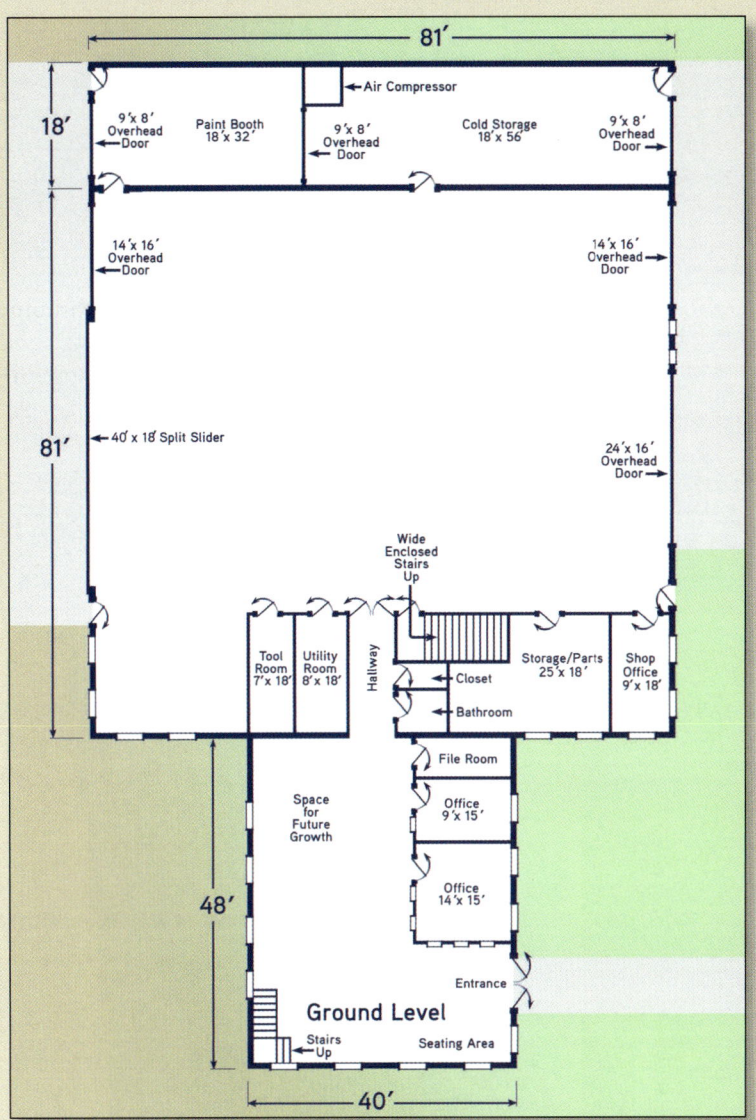

81'

18'

9'x 8' Overhead Door
Paint Booth 18' x 32'
Air Compressor
9'x 8' Overhead Door
Cold Storage 18' x 56'
9'x 8' Overhead Door

14'x 16' Overhead Door
14'x 16' Overhead Door

81'

← 40' x 18' Split Slider

24'x 16' Overhead Door →

Wide Enclosed Stairs Up

Tool Room 7'x 18'
Utility Room 8'x 18'
Hallway
Closet
Bathroom
Storage/Parts 25'x 18'
Shop Office 9'x 18'

File Room
Office 9'x 15'
Office 14'x 15'

48'

Space for Future Growth

Entrance

Ground Level

Stairs Up
Seating Area

40'

left: The first floor of the Schaffer complex comprises a cold storage area and paint booth (under an 18×81-foot lean-to structure attached to the shop), the work floor of the shop, nearby storage rooms, shop office, and finally the attached office area (at bottom of the illustration).

below: The second floor of the complex houses shop storage, bathroom, kitchen, and the office loft. That loft provides for future expansion of the office. The office reception area (at bottom right of the illustration) is open to the ceiling.

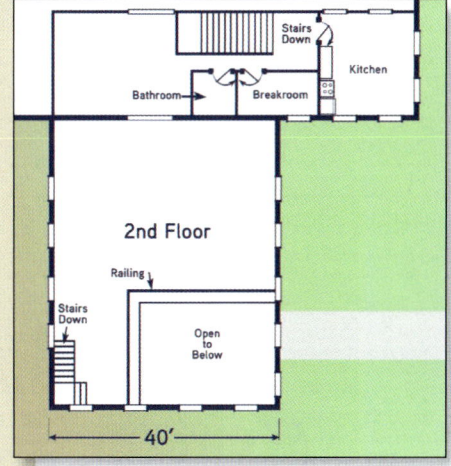

Stairs Down
Kitchen
Bathroom
Breakroom

2nd Floor

Railing

Stairs Down

Open to Below

40'

above and top right: The shop's massive 63×81-foot work floor can accommodate a wide variety of large equipment at once. Patrick Schaffer positioned multiple doors on both ends of the shop floor to allow equipment to readily flow through the shop. This work area is kept unobstructed since all tools and supplies are stored in multiple storage rooms (accessed through doors to the left of the combine).

bottom right: Stationary tools are housed in an 18×24-foot work space that also serves as the shop's machining area.

Grand Champ Shop

The northwest corner of the Smith shop houses the operation's office complex. The entrance location was chosen since it's close to existing weight scales.

Spacious shop plan and office complex for larger machinery

By Dave Mowitz

The Smith family faced a dilemma that's all too common among growing farms: More acres demanded wider equipment to keep pace with planting and harvest.

The family's old shop – once ideal for eight-row heads, 16-row planters, and straight trucks – couldn't house larger machines.

So the Rochester, Indiana, farm team of Dave and Kevin Smith tapped into the expertise of their father, Dale, and a team of employees to design a facility for the future – a shop where an entire fleet of machinery could be tended to under one roof at one time.

The result was a facility that can readily accommodate a 36-row planter – unfolded – with room left over for a semitrailer and a couple of tractors or pickups. Yet the Smith shop is more than large, sitting on an 88×96-foot pad topped off with 20-foot sidewalls and a custom-built clean-span ceiling. Their layout provides for free-flowing repair and maintenance bays, ample parts and supply storage, and an adjoining office complex.

"When specing out the shop, we opted for 20-foot sidewalls since who knows what the future will bring with larger equipment," Kevin Smith explains. "That height allows us to unfold toolbars in the shop plus provide room to a 20-foot-tall HydroSwing door."

OPENINGS INTO THE SHOP

Door placement was a crucial element in the Smiths' shop layout. The family placed the 40-foot-wide HydroSwing door at the east end of the shop floor. This door provides access to the shop's main service bay,

The east side of the shop provides access to a massive service bay (shown below) through a 40-foot-wide HydroSwing door. The 16-foot door to the right opens into a wash bay.

Photographs: Ron Van Zee

The favorite feature of their shop is a 35-foot-long pit that offers access to the axles of a semitrailer, explains Dale Smith. "At 44 inches wide it also offers space for two people to work in the pit at the same time," he says.

The Smiths had 36-inch-deep workbenches custom-made from steel plate. Tools hang on slotted-board hangers.

which is over 80 feet wide and nearly 70 feet deep.

Located next to that main opening is a 16-foot overhead door that "offers quick access into the shop for vehicles and tractors," Kevin says. "In winter, this area acts as our wash bay because it has a floor drain. In the summer, the outdoor concrete pad in front of the door is the wash area."

A third opening was located in the southwest corner of the shop. This 24-foot-wide overhead door provides access to a service pit running along the west side of the structure.

SERVICE ENTIRE LENGTH OF A SEMI

This pit was designed to provide access to all the axles of a semitrailer tractor. "That pit is one of the best investments we made," Kevin says. "Its 5-foot depth allows us to work underneath vehicles without hitting our heads. Yet, we can reach nearly every grease point on a truck or trailer. With that access, we're also able to adjust brakes or tend to other truck maintenance chores."

Rather than incorporate their office inside the shop's main floor area, the Smiths opted to construct a 20-foot-wide addition on the north side of the structure. This addition runs the full 96-foot length of the overall structure. Inside that addition resides a complex of rooms that includes four offices, a spacious combination break room and kitchen, and bathroom. A parts storage bay (shown at right) occupies the 30-foot-long area at the east end of the addition. "We could have had a separate office building," Kevin says, "but the shop has become the hub of our operation and central meeting place for the family and employees." ∎

A 20x30-foot parts storage area was positioned off the shop's combination wash and service bay next to the 16-foot-wide door. A staircase (located out of the picture to the right) leads to a 20x60-foot storage loft, which is totally enclosed except for a forklift access opening into the main shop. "Walling off the storage area not only keeps items out of sight, but also prevents dust and smoke from accumulating up there," Kevin Smith says.

New Shop, New Business

A complex of a shop and attached office provides base for diversified farm operation

By John Dietz

Pat Muller designed his shop complex by attaching a 24×42-foot office on the side of the 66×75-foot main shop structure. Recently he poured a 50×75-foot concrete slab behind the shop to serve as an unloading space and to provide temporary outdoor storage. This area comes in handy for a sideline business Muller developed assembling implements for equipment dealers.

A crowded old shop can be frustrating. But for Pat Muller, it was the starting point for a new business. Frustration with his old structure pushed the Hillsboro, North Dakota, farmer to invest in a new shop, which, in turn, inspired him to add a number of sideline businesses to his operation.

"I had to have something bigger," he recalls. "In this shop, we can work on four to five projects at a time."

When they're not farming 2,500 acres of corn and soybeans, Muller, sons Eric and Brandon, brother Ron, and employee Dan Linden keep constantly moving by custom spraying, delivering seed, running a planter and sprayer parts mail-order business, and assembling implements for local dealers.

That diversification would not have been possible, Muller believes, were it not for the investment in a 66×75-foot shop and attached 24×42-foot office over a decade ago. Within six years of its construction, Muller estimates, the complex paid for itself.

That was no small accomplishment since one of the shop's major costs was the installation of an expensive geothermal heating and cooling system. This system consists of a closed loop of plumbing that circulates a mixture of isopropyl and water through some 22 wells drilled to a depth of around 120 feet. That solution circulates through two exchangers rated at 5 tons and 6 tons in operation capacity, respectively.

GEOTHERMAL PAYOFF

These exchangers operate to heat or cool air circulated through the structure. "When the contractor installed the air conditioning side of the system, I thought I'd never use it," Muller recalls. "Today, we use the air conditioning all the time to keep the work environment in the shop comfortable. As a result, we get more work done."

Comfort aside, the real payoff for the geothermal system is in operating costs. "With it I can heat the shop and office for pennies a day," Muller points out. "And that's even when it's -30°F. and windy!"

A good example of this was the winter of 1996-1997, which was particularly long and cold. In the spring, Muller was comparing notes with a local dealer who was heating his slightly smaller shop with propane. "He spent $1,500 a month on heat, while I was going through a little over $100," Muller recalls. "That's the cost of heating not only the shop, but also the office and part of our nearby house!"

The shop's balcony is 4½ feet wide and runs the full length of the south wall of Pat Muller's shop.

A ventilation hood is located over Pat Muller's welding area to remove toxic fumes and to keep the shop free of dust.

Pat Muller's desk is located on one side of the mostly open 34×42-foot office space.

When the shop was erected, Muller opted for a sturdy design featuring three 2×8-inch laminated columns spaced at 7½-foot intervals. The columns are anchored more than 4 feet deep.

The structure was then insulated to provide R-50 protection in the ceiling (with 15 inches of blown insulation) and R-20 on sidewalls (using fiberglass batts).

To finish out the structure, Muller covered the shop's walls with 26-gauge steel panels. The office was complete using drywall.

Shop lighting consists of a mixture of fluorescent fixtures along the sidewalls to provide spot lighting for detailed work. The main light source, however, is high-pressure sodium lights. Muller prefers such lights in the ceiling for their very even illumination that casts few shadows.

PASS-THROUGH DOOR ACCESS

Access to the shop is provided by two doors. The main entry is a 30-foot-wide door. At the back of the shop is a 22-foot-wide door. Each overhead door has an electric opener.

Strategically located near the back door is a steam and hot water pressure washer. The addition of a concrete pad outside of this door makes a ready wash area. Also located in the same area as the pressure washer is the shop's oil bay complete with bulk oil and antifreeze storage tanks.

Positioned at the other end of the shop and near the front door is the shop's welding center. Muller took extra steps to install a ventilation hood with a massive welding table.

FORKLIFT A MUST

Along the wall near the welding area are a hydraulic press, metal band saw, and lathe. And the oomph to lift and locate machinery being assembled is provided by an 8,000-pound, three-stage forklift equipped with a side-deliver feature. "This is a must for our shop," Muller adds. "Our first forklift was a diesel. We later switched to a lift that was powered with propane, which is much cleaner."

OFFICE AREA LEFT OPEN

Muller avoided partitioning his office structure. The area is open except for a utility room and washroom. Two offices, a lounge area, kitchenette, library, and bins for their mail-order parts business are located around the room.

"This spaciousness will let us change the office layout or expand somewhere, such as the mail-order area, if we need to in the future," Muller explains. ∎

Designing an Expanding Shop

Joel (front) and Dolf Jakobs opted to erect a 66-foot-wide structure that provides work area to service or repair two major pieces of machinery.

The key is the strategic location of specialized work centers

By Dave Mowitz, Machinery Editor

Seems like any discussion you have with farmers about their shops ends with them uttering, "I wish I would have built larger."

Yet upon questioning, these farmers often reveal that they did, in fact, give ample consideration to anticipated repair chores and future growth in machinery size before building. As it turns out, the main reason their shops seem to shrink after a few years of use is because their appreciation for modern facilities has grown.

For example, prior to erecting the shop, they may have never tackled a maintenance chore larger than changing oil. In a new structure equipped with an overhead crane and expansive workbenches, they find it's easier to take on an engine overhaul.

In the old days, welding jobs usually consisted of spot repairs. But in a new shop, these farmers find they have room to spread out and create their own implements from scratch.

While it's not always possible to increase the structural size of existing shops, there are many possibilities for making them expandable through strategic location of work centers, say Vern Hofman and Kenneth Hellevang. The North Dakota State University engineers first devised the concept of creating separate repair, fabrication, and service bays as part of a general shop plan in their publication, *Planning Farm Shops*.

The heart of an expandable shop is strategic placement of work centers, or

bays. These locations are surrounded by supporting service centers. Hofman's and Hellevang's research of farm shops combined with 20 years of shop coverage in *Successful Farming* magazine reveal that most successful shops encompass at least two of these types of work centers:

■ Major repair and/or fabrication bay
■ General service or maintenance bay
■ Combinations of both

These major work centers are surrounded by supporting service centers, which might include one of these:

■ Welding center
■ Machining (metalworking) center
■ Wash area
■ Lubricant storage
■ Storage loft
■ Workbench repair area
■ Tire repair center
■ Shop office

TOP SHOP TIP:

20-POUND FIRE EXTINGUISHER A MUST

Time was when a 5-pound fire extinguisher was all the protection needed for a farm shop. Not anymore. "A 5-pound extinguisher provides just 30 to 40 seconds of firefighting ability," explains George Maher of North Dakota State University.

Increased welding activity combined with the increased storage of highly flammable items requires any shop to have at least one 20- to 25-pound extinguisher, Maher urges.

Place 10-pound extinguishers near high-risk areas, such as lubricant storage, the welding center, and workbenches. The best extinguisher for shops is a Class ABC unit, Maher adds.

As in real estate, the key to making a shop internally expandable is location, location, location. In the case of shops, this involves strategically positioning work bays with service centers in conjunction with myriad structural features like doors, floor drains, hoists, lighting, electrical outlets, etc. "You can do this keeping your specific operation's needs in mind," Hofman says. "A livestock operation has different repair or fabrication needs than, say, a straight crop operation."

Generally, a shop's major repair and/or fabrication bay should be located in the largest area of the shop. This bay should also be positioned in front of a shop's main entrance. This makes certain that a repair or fabrication bay can readily accommodate a farm's largest equipment.

SUPPORTING SERVICES

Supporting service centers, like machining or welding areas, are best located near a repair or fabrication bay. Special equipment such as cranes, floor anchors, or exhaust fans are best positioned in this work bay, too.

If a separate service bay is created to accommodate day-to-day maintenance chores, like changing engine oil or greasing equipment, then it is best to locate this work center next to your shop's secondary door.

Likewise, position service centers such as a tire repair center, service pit, and lubricant and related supply storage (filters, grease, etc.) near a service bay. "For example, you wouldn't want to locate lubricant storage in a fabri-cation bay if you are changing oil in the service bay at the opposite end of the shop," Hofman points out.

Wash or paint bays present a unique challenge. Such work bays may require their own separate area in a shop to prevent water or paint from being sprayed on equipment under repair. Or, they can be created as part of an existing bay through the use of curtains or a collapsible partition.

TOP SHOP TIP

WIRE FOR THE FUTURE

At the very minimum, any shop should be wired with 200-amp, 240-volt electrical service. Today, however, most farm facilities need 300-amp or even larger service.

As a rule, install 120-volt duplex outlets every 4 feet along workbenches and every 10 feet along walls. Ground fault interrupters (GFIs) are needed on all single-phase 15- and 20-amp circuits, and they're a must in outlets in wet areas or outdoors.

Install separate branch circuits (wired to their own breakers) for outlets serving ½-hp. or larger motors. An individual circuit should serve no more than three ⅓-hp., two ½-hp., or one 1-hp. motor. Welders must have their own 50-amp, 240-volt circuit.

For more information, go to the National Food & Energy Council at www.nfec.org.

TOP SHOP TIP

VENTILATE WELDING FUMES

Ventilating a shop was once considered a luxury. But it's a necessity, particularly when welding stainless steel, cadmium- or lead-coated steel or nickel, chrome, zinc, or copper. Fumes from these metals are considerably more toxic than those encountered with mild steel, warns the National Institute for Occupational Safety and Health.

Medical research indicates welders have a 40% increase in odds of developing lung cancer. Other ailments from long-term exposure to welding fumes include asthma, emphysema, chronic bronchitis, and fibrosis of the lung. A common reaction to metal fumes – overexposure to zinc oxide fumes in particular – is flu-like symptoms.

Ventilation is key to avoid breathing such fumes. Timothy Lawrence at Ohio State University says ventilation is sufficient in these settings:

- The welding area contains at least 10,000 cubic feet and the ceiling height is not less than 16'.
- Shop cross ventilation is not blocked by partitions, equipment, nor other barriers.
- Welding is not performed in a confined space.

If these requirements aren't met, then add ventilating equipment that exhausts at least 2,000 cfm of air, except where local exhaust hoods or booths or air line respirators are used, Lawrence adds.

The best kind of ventilation removes fumes before they pass by the welder's face. This can be supplied by a flexible hose (6" to 8" in diameter) equipped with a hood. Place the hose and hood 6" to 9" from the welding work. The fan servicing that hose should supply 550 to 750 cfm of exhaust air movement.

Viable options to hose ventilation are a supplied air-type respirator or a respirator specially designed to filter the specific metal fume.

Also, when outdoors, try to stand upwind from welding fumes.

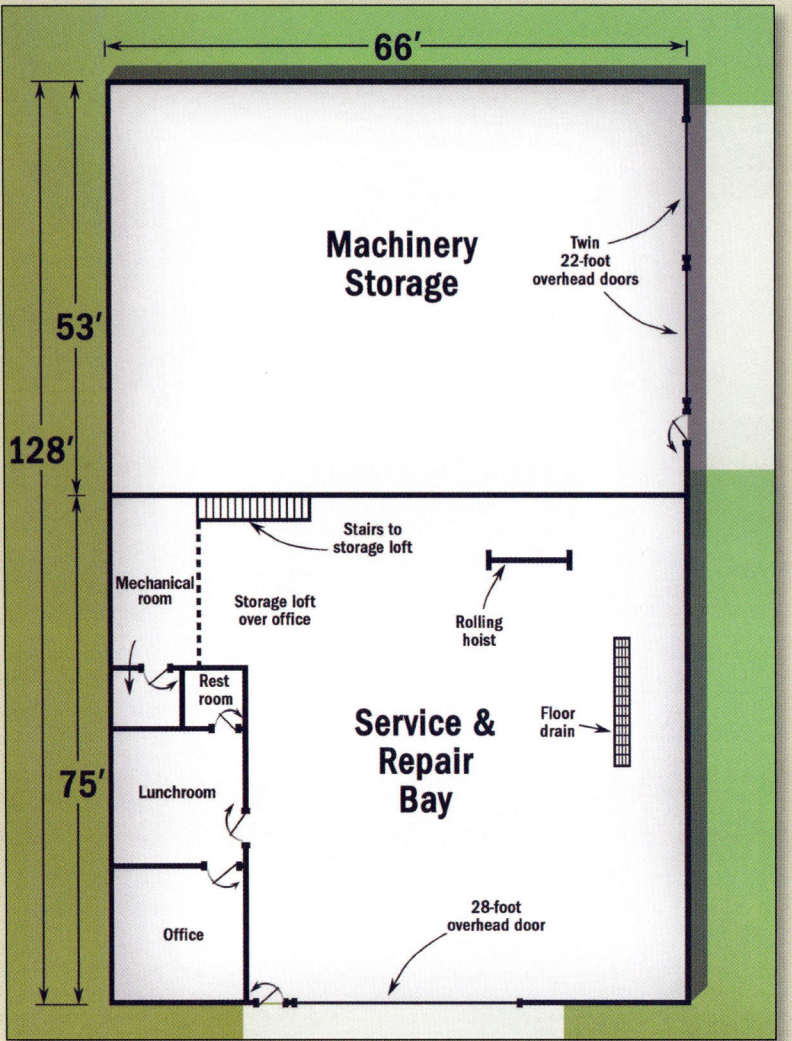

66'

Machinery Storage

Twin 22-foot overhead doors

53'

128'

Stairs to storage loft

Mechanical room

Storage loft over office

Rolling hoist

75'

Rest room

Service & Repair Bay

Floor drain

Lunchroom

Office

28-foot overhead door

The Jakobses use a single building to serve as shop and machinery storage.

A rolling crane can service any location on the Jakobses' shop floor. The shop's major workbench is located under a storage loft.

Whatever its location, a wash bay will need to be supported by a floor drain and GFI-protected electrical outlets. Of course, an exhaust fan and paint supply storage is best located near a paint bay.

"Logic has a lot to do with where work centers and tools are positioned," Hofman says. And logic is certainly at work in the five farm shops featured.

1. TRADITIONAL LAYOUT

J&J Farms of Milledgeville, Illinois, took a traditional approach to layout by carving out a combination service and repair bay in the front of their 66-foot-wide structure. This bay, accessed by a 28-foot-wide door, accommodates the livestock operation's many repair chores. The key to its success is the ample width of the bay, which allows Dolf Jakobs and his son, Joel, to park major equipment like combines and tractors side by side for major repair or simple service. A welding center and related metalworking tools are logically positioned near the repair bay for easy access.

The Jakobses located a 15×40-foot office and lunchroom in one corner of the front of the shop. As is common practice, the area over the office was converted into a storage loft. The Jakobses then extended the loft along the side of the shop. This then created a logical location for a workbench and tool storage under the loft.

To save on construction expense, the Jakobses partitioned off the rear 53 feet of the building for machinery storage. Some other features that the Jakobses designed into their structure include in-floor radiant heat, a homebuilt portable hoist, and perforated steel walls "that help absorb noise," Joel explains.

2. IN-LINE BAYS

Ron and Greg Otto's approach to their layout was to position a major repair and fabrication bay at one end of their 50×102-foot structure that is accessible by a 35-foot-wide door. The Lester Prairie, Minnesota, farm team mounted a homemade 20-foot jib (or swinging) crane to one side of this primary entrance. Welding and metalworking tools are parked at the base of the hoist. The Ottos installed a grid of floor anchors in the center of the fabrication bay that, as a result, created a 16×32-foot construction area.

At the other end of their shop is the service bay, which is equipped with a service pit and lubricant storage. Access to this separate work center is provided by a 16-foot side door.

Other features of the Ottos' shop design include air and electrical outlets spaced every 10 to 12 feet around the entire structure, and fluorescent lights that are wired to separate switches. This way, the repair-fabrication bay can be illuminated apart from the service bay.

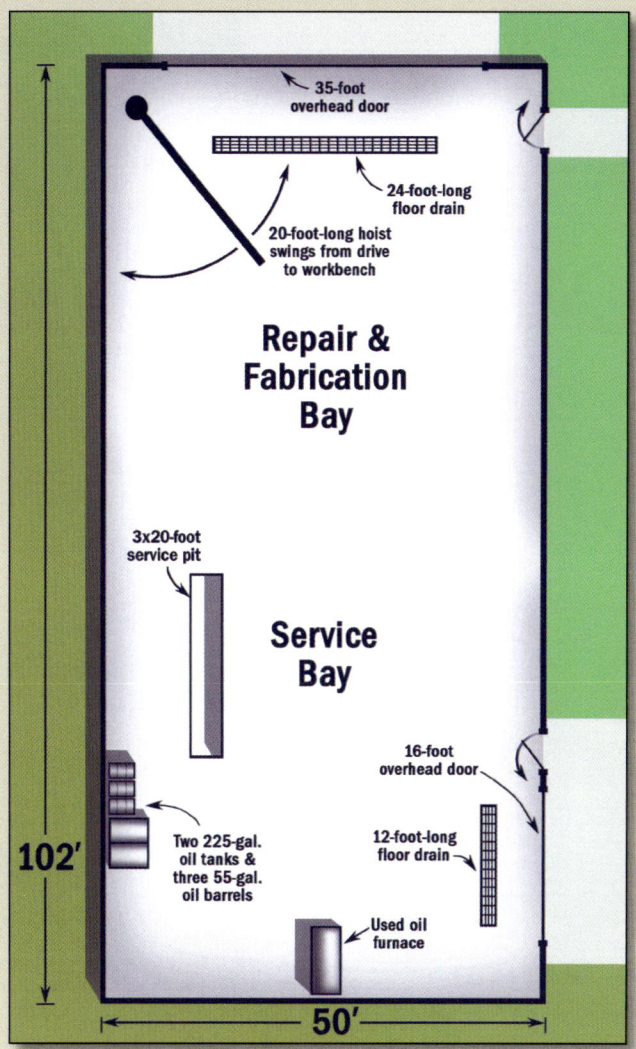

The Ottos provided a separate door to access their service bay.

TOP SHOP TIP

INSULATE TO THE MAX

If you're building a new shop or haven't insulated your existing shop, then figure on installing a minimum of R-15 to R-20 insulation in walls, R-25 to R-30 in the ceiling, and R-10 in your doors. A vapor barrier of 6-mil polyethylene should be installed between insulation and the inside wall panel. Insulate the foundation (new construction) with a minimum of 2 inches of polystyrene insulation (R-10 to R-12) to a depth of 24 inches.

Heidi, Greg, and Ron Otto located a 20-foot-long, 2-ton pivoting hoist near their shop's main entrance to serve their main repair and fabrication bay. It also swings outdoors for working on wide implements.

A work pit is located to one side of a service bay and near an ample 800-plus-gallon lubricant storage center. Used oil is pumped from the pit to an outdoor storage tank where it feeds the shop's used oil furnace.

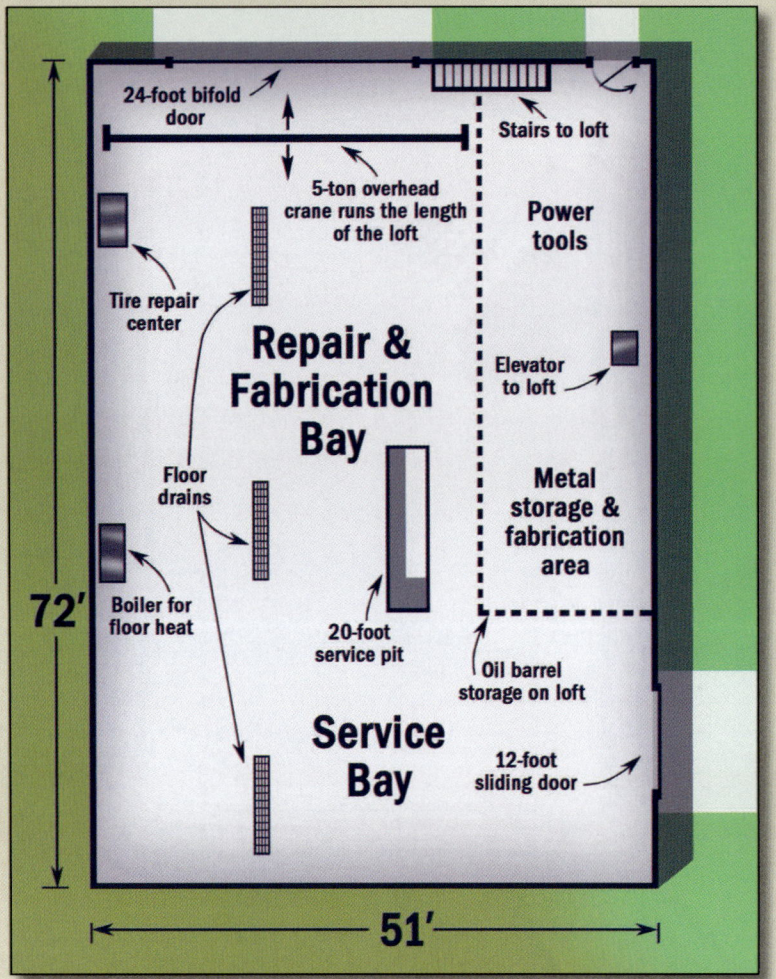

24-foot bifold door

Stairs to loft

5-ton overhead crane runs the length of the loft

Power tools

Tire repair center

Repair & Fabrication Bay

Elevator to loft

Floor drains

Metal storage & fabrication area

72′

Boiler for floor heat

20-foot service pit

Oil barrel storage on loft

Service Bay

12-foot sliding door

51′

The Leonards located all their major metalworking tools near the repair and fabrication bay. A pit and overhead oil storage are near the service bay.

3. SHOP LAYOUT CREATED FOR MAJOR FABRICATION

Fabricating machinery is a major occupation of Jim Leonard and his sons, Dean and Duane. So the Morgantown, Indiana, farm team put a repair and fabrication bay in their 51×72-foot shop. That work center is positioned in front of the shop's 24-foot-wide bifold door and under a 5-ton overhead trolley crane constructed to service the 35×48-foot bay.

An extensive workbench and machining center, complete with metal storage, is positioned along the fabrication bay and under a storage loft. This machining area is home to a variety of welding equipment, drill presses, a metal band saw, hydraulic press, and pedestal grinders. Metal storage racks are located at the end of the storage and fabrication area.

A unique feature of the Leonards' storage loft is a 3×6-foot homebuilt elevator the Leonards built to lift parts and supplies to the loft. They also added a staircase to the loft located by the shop's main door.

A separate service bay sits at the opposite end of this shop and is ac-

The overhead crane runs over the width of their fabrication bay, which is next to a machining area.

Jim Leonard and sons Dean and Duane are master fabricators and make ready use of a 5-ton trolley crane. That crane, which provides two chain hoists, was positioned so as to provide 14 feet of clearance.

cessed by its own 12-foot door. This bay was equipped with a 20-foot-long service pit as well as overhead lubricant storage in the loft.

4. SEPARATE WASH AREA

Brit Liljedahl's shop layout provides insight on how to provide room for a wash area. The shop, designed with help from brother Mike and their late father, Lenus, encompasses a wash center located along one side of the shop and in front of a 20-foot door. This area is equipped with a workbench, hydrant, sink, and 3×5-foot pit.

Primary service and repair activities take place in a bay located behind the shop's main 30-foot bifold door. This work center is serviced by a nearby workbench as well as parts and lubricant storage areas (in an overhead loft).

The Missouri Valley, Iowa, shop also carves out room for a welding center in one corner of the structure. This miniature fabrication bay can be accessed by either of the shop's two doors to service two repair jobs at once.

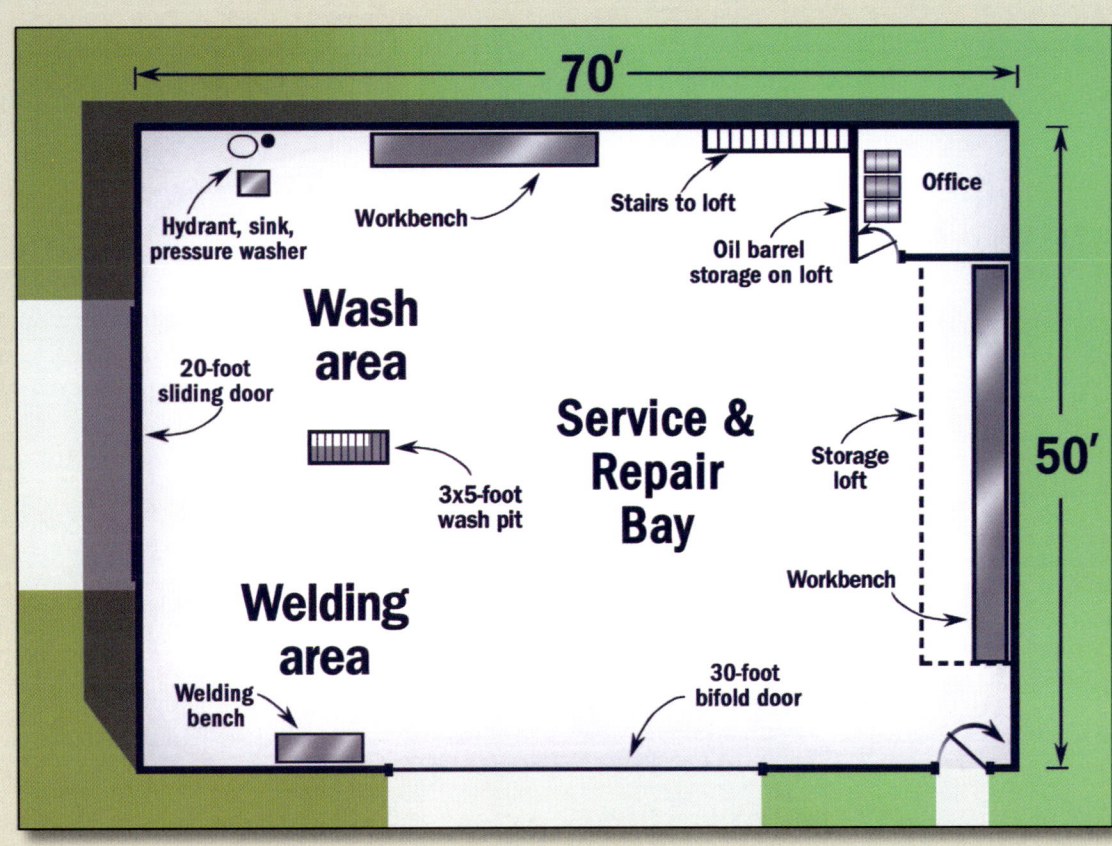

The Liljedahls carved out three separate work centers from their structure and surrounded the bays with supporting services.

Brit (left) and Mike Liljedahl, with their late father, Lenus, designed their operation's 50×70-foot shop.

The Liljedahls created a wash area that is accessed by a 20-foot sliding door on one side of their shop. A welding center is positioned opposite of the wash area and near the shop's main 30-foot door.

The Liljedahls' shop office is positioned under a storage loft. This loft extends along the side of the shop and covers a major workbench.

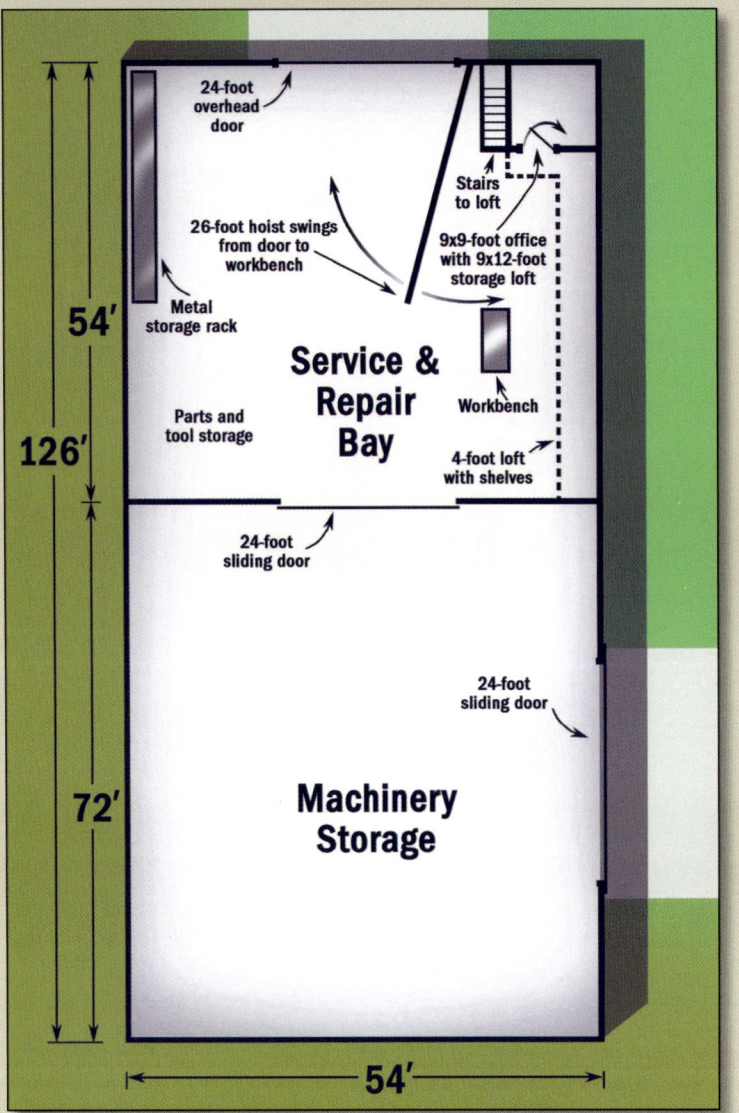

24-foot
overhead
door

Stairs
to loft

26-foot hoist swings
from door to
workbench

9x9-foot office
with 9x12-foot
storage loft

Metal
storage rack

**Service &
Repair
Bay**

54'

Parts and
tool storage

Workbench

126'

4-foot loft
with shelves

24-foot
sliding door

24-foot
sliding door

72'

**Machinery
Storage**

54'

Robert Kuesel segmented a 126-foot-long structure into separate shop and storage areas with an access door.

5. SHOP, STORAGE ALL IN ONE

A unique swinging crane that also slides along a rail is the most unique feature in Robert Kuesel's shop. But, the best part of the Napoleon, Ohio, farmer's shop is a 24-foot door he put between his shop and machinery storage. This door lets Kuesel move machinery to and from the shop from storage without going outdoors during cold winter months. And, access to the 54×72-foot area gives him storage for oversize metal or large parts.

Otherwise, a wealth of storage is found in a loft that starts out over Kuesel's 9×9-foot office and extends along the side of the shop via a 4-foot-wide walkway complete with shelves. The shop's workbench area is located under this walkway. ∎

Robert and Nancy Kuesel have discovered their structure acts as their operation's headquarters, thanks to a shop office, as well as shop.

Kuesel's main repair and fabrication area is serviced with a unique crane that swings out over work. The crane's pivoting point slides down a rail to expand its working reach.

Multipurpose Building Houses a Shop

Innovative use of existing components allowed the Shryocks to re-create the look of a classic red barn with all the modern conveniences of new construction.

Maze visitors and machinery repair all in one structure

By Dave Mowitz

This is certainly not your typical farm shop. The exterior of Shryock's Callaway Farms headquarter looks vintage red barn. Yet the structure is modern in every way, built entirely of components from Lester Buildings.

Enter the front of the building and you walk into knotty pine-lined walls reaching to the peak of the roof – a perfect setting to welcome the over 15,000 people who visit this family farm's corn maze each fall.

But the back of the 68×80-foot building is all shop cleverly laid out to accommodate a wide variety of repair and maintenance chores at the same time. "Designing the building to provide these uses justified the investment," says Denny Shryock, who farms with brother Joe and their respective sons, Mike and Brett, near Fulton, Missouri. ∎

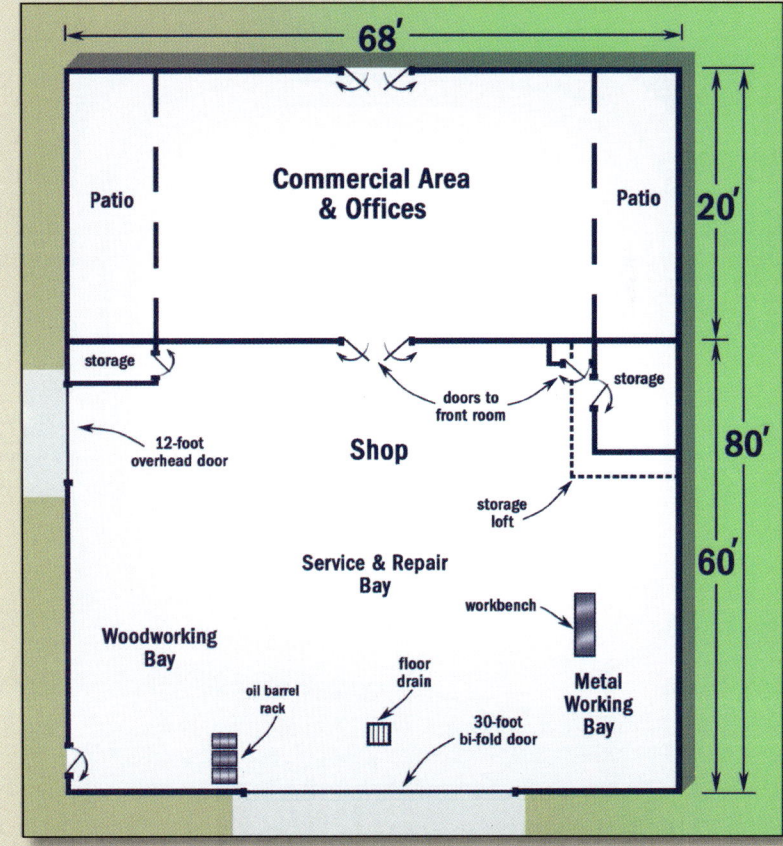

The Shryocks crafted their farm's headquarters to suit a growing sideline business while still servicing their operation's machinery. The front of the building houses an open area for customers coming off the parking lot. The rear of the structure serves as the shop divided into various work bays. The 30-foot door on the end accommodates large machinery for major repair and fabrication work, while a 12-foot side door provides access for light maintenance chores.

Photograph: Doug Hetherington,   Floor plan: Paul Bridgford

Top Shops® Design Champ

The entire floor space of the Minors' 80×20-foot shop is accessible by four overhead doors.

Spaciousness, great lighting, and a wealth of features win first place for this shop

By Dave Mowitz, Machinery Editor

Years of working in cramped conditions in a drafty wooden shed had Brad and Terri Minor yearning for a new shop. "It was a fine building for its time," Brad recalls. "But our equipment outgrew that shop."

All that time, Brad was making plans for a new shop to accommodate the family's growing operation near Rutland, South Dakota. The family kept a folder with the best shop ideas they saw in farm publications. "That was my design book," Brad explains. "My family and employee, Mike Carroll, were the design team. We put ideas down on paper until a plan was created that we all liked."

That planning became an 80×126-foot structure with many features that won the Minors the Best Shop Design category in *Successful Farming* magazine's Top Shops® contest.

LOST IN SPACE

What strikes you when first entering the Minors' shop is its spaciousness. The structure has 18-foot sidewalls and a raised truss with a 22-foot peak. "This gives us 18 feet of clearance for the overhead doors," Brad says.

This cavernous interior readily houses the farm's largest equipment. The family had considered walling off the end of the shop for cold storage. "We held off on that, and I am glad we did," Brad adds. "We've since found we needed that extra space. We discovered that a nicer shop encourages us to do more repairs and fabrication at home."

The shop's spaciousness is made more useful by great illumination. The

The Minor shop design team includes (from left) Dwaine and Helen Minor (Brad's parents), Brad and Terri Minor, son Mike, 11, and employee Matt Carroll.

ceiling is finished off with white metal sheeting and then wired with 34 high-pressure sodium lights.

FEATURE ATTRACTIONS

The Minors are still adding finishing touches to their shop. Some of the outstanding features already in

Photographs: Ron Van Zee

place include:

■ **An automotive hoist and a drive-over work pit located near lubricant storage.** "The pit has fluorescent lights and electrical plug-ins," says employee Matt Carroll. "It also has a tank for waste oil. We installed a line under the floor that leads to an outdoor waste oil tank. We use air pressure to blow the oil to the outside tank."

■ **A room on the storage loft that houses the air compressor and power washer.** "It is insulated with 6-inch batt insulation to absorb the noise," explains 11-year-old Mike Minor. The compressor and washer are on rubber mounts attached to a ¾-inch plywood false floor.

Dual front doors share a common track post. That post slides to the side to create a 45-foot-wide main opening into the shop, allowing a combine with a head to enter.

Styrofoam board, 2 inches thick, is sandwiched between that false floor and the loft's ¾-inch plywood deck to absorb vibration.

■ **A 12×40-foot office lined with stained wood siding.** The shop's bathroom, electrical panel, and plumbing for the floor heat (a waste-oil boiler is located on the loft) are housed in a separate room. ■

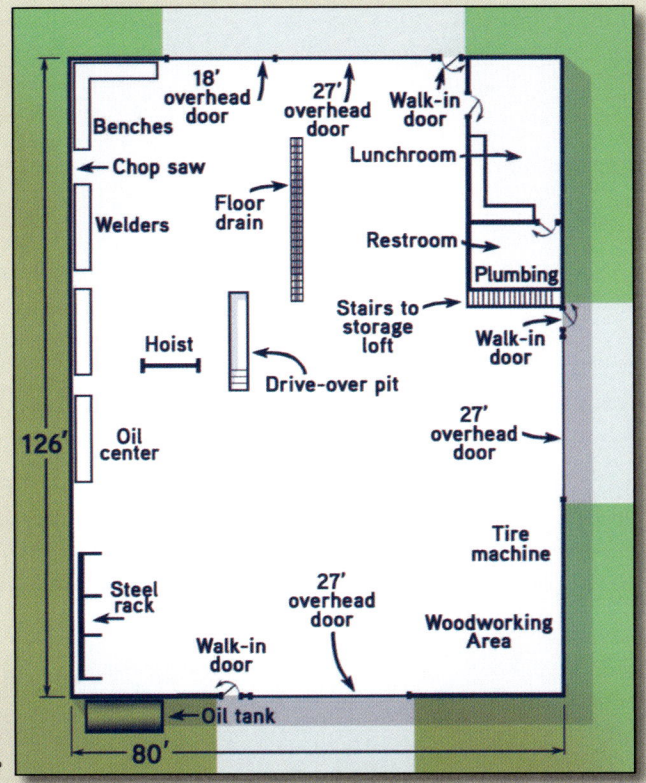

The three entrances into the Minors' shop combine to create three work bays. That allows multiple vehicles and implements to be serviced or repaired at the same time. The Minors positioned an automotive hoist and drive-over pit so vehicles can approach from either the front or rear doors.

Illustration: Paul Bridgford

top: Matt Carroll works in the oil pit,
bottom: Son Mike suggested the storage loft.

Double-Duty Shop

Innovative equipment makes Steve Bughi's shop far more versatile

By Denny Eilers

Steve Bughi has found that his 30-foot-long workbench is one of his shop's most important assets. The roomy workspace also offers invaluable storage shelves located under the work platform.

Growing up, Steve Bughi couldn't wait to own a shop. Once that dream came true, he realized he could use the shop for much more than changing oil and rotating tires.

Each Thanksgiving Bughi and his wife, Patty, host family and friends for an old-fashioned hog butchering – making sausage, bacon, and smoked hams. These gatherings take place in the Bughis' 40×60-foot shop located near Rosalia, Washington.

This versatile shop contains not only his tools and projects, but also the memories of family events.

Bughi looked at several options before deciding on a 2×6-inch frame for his structure. These dimensions allowed for needed room to accommodate insulation and drywall on the inside of the building. He says insulation is important – especially in a shop with 16-foot ceilings – because it maintains heat in winter and stays cooler in summer.

SHOP LAYOUT

One side of the building is arranged to handle welding and the other side has a 30-foot-long workbench with shelving for storage. A covered walkway connects Bughi's shop to a shed where he stores extra parts. "I try not to store too much stuff in the shop. That way I have more room to work," he explains.

While working on equipment, Bughi uses a roll-around tool chest with hand tools and wrenches. A separate shelf area stores saws, blowers, and additional hand tools.

Bughi recently added a specially designed overhead hoist to improve efficiency. With its A-frame shape and 4-ton lift capacity, he can quickly pull transmissions or engines, and he's even used it to lift cabs from tractors. The hoist is made of tubular steel and has four wheels that can be moved around the shop and locked into position.

"It's tall enough to go over the top of tractors or combines," Bughi says.

Bughi's dreams came true when his own shop was built. It is used for equipment maintenance and much more.

Bughi dumps used engine oil into a home-built collector, which is designed to sift out dirt and chunks. The collector has proven to be vital, because Bughi recycles all the engine oil consumed in his shop.

Steve Bughi's insulated shop keeps it the right temperature whether he is working in the office or on equipment.

Another handy unit is a small electric forklift with a 3,000-pound capacity that he uses for changing tires and for doing other small lifting chores.

"It's indispensable for pulling off tires," he says. "It can easily pick up a tire full of fluid from one of the big tractors."

Bughi also built a freestanding dolly that's kept full of wood blocks of various sizes – railroad tiles, 4×6s, 2×4s – all cut to about 3-foot lengths.

"When you jack something up, you roll the dolly over and have plenty of blocks to put under it," he says.

SHOP EXTRAS

Bughi also created an easy way to dispose of engine oil. "I took an old pickup toolbox, put legs on it, and then placed a fine mesh screen over the top," he explains. "When I change oil, I pour the used oil into this container and the screen takes out the dirt and chunks. Then I set the filters on the screen mesh and let them drain."

A small pump moves the oil from the container that's inside the shop into a closed 350-gallon storage tank outside the shop. "When the outside tank is full, I call and have a person come out and pick up the waste oil. This has been a very handy way to take care of a really messy problem," he says.

Bughi states the one thing he would do differently were he to build his shop today is to put in a floor drain so he can wash equipment inside his shop.

But when it comes to using the shop for Thanksgiving traditions and celebrating his daughters' graduations, he smiles and says, "For that, I wouldn't change a thing." ■

Literally a Shop on Wheels

The van body of a used truck provides the base for a true shop on wheels

By Dave Mowitz, Machinery Editor

The van body with wooden floors and wood-reinforced walls proves ideal for wiring and securing tools. "The only thing I wish the truck had was a crew cab so that more than two people could ride to the field in it," Marvin says.

When Marvin and Matt Mechtel were first envisioning their dream service truck, they did some out-of-the-box thinking. The result was literally a shop on wheels that is *in* a box.

As well equipped as many small shops, the only thing the Mechtels' service truck lacks is much floor space. But it is ready to go to work servicing most any repair and maintenance chore in the field while providing an enclosed environment to work in. "We tried to stock the truck with practically anything you'd use for field repairs. Plus, all the tools and other components like the compressor and generator set are out of the weather," says Marvin.

For their ingenuity, the father-and-son duo won top honors for the Best Service Vehicle in *Successful Farming* magazine's Top Shops® Contest.

VAN BODY BOX IS IDEAL

The home for the portable shop is a 1995 Freightliner FL 70 truck with a 24-foot van body that features a roll-up, full-width rear door and hydraulic liftgate. "We found the truck in Kansas," Marvin recalls. "It had less than 200,000 miles on it, was in excellent shape, and was priced well under $10,000."

The Freightliner and its body needed very little modification for the new role as a service vehicle. Once they got it home to their farm near Page, North Dakota, the Mechtels set about outfitting the body.

One of the first items installed was a custom-built, 750-plus-gallon fuel

"This truck is stuffed full of more tools than most farm shops," laughs Marvin Mechtel, who farms with his son, Matt. "Actually, I think it has more tools than the shop I had when I first started farming."

tank mounted in the very front of the van body for good weight distribution. "It is plumbed to a 30-gallon-per-minute, 220-volt Tuthill pump feeding 50-plus feet of hose that is wound around an electric hose reel salvaged, along with its meter, off of an old fuel delivery truck," Marvin recalls. "All diesel fuel passes through a Centek water and particle filtration system."

Fuel hoses pass through a side door on the curbside of the truck.

For easy access to the box, the Mechtels made a lightweight but stout-as-concrete staircase that when unfolded provides a sturdy handrail.

Photographs: Ron Van Zee

THERE'S A LOGIC TO TOOL LOCATIONS

Positioning the truck's host of major tools took careful consideration. For example, the stationary generator set is at the front of the box near a fuel source.

Power from the generator is distributed along both sides of the body's walls for easy access.

"The box has a number of high-wattage light fixtures to illuminate the interior for night work," Marvin points out. "And we also mounted two 500-watt halogen lights outside on the curbside of the truck for doing outdoor repairs when it's dark."

Next to the generator set and mounted to the wooden floor or reinforced walls along the left wall (as viewed from the back door) is a stacked toolbox and steel welding bench equipped with a vise and drill stand. Next to the stand is a Millermatic 250 wire welder.

"The welder is by the back door so its cables reach the bench as well as outdoors," Marvin says. "Or we can roll it out on the liftgate and to the work site."

The last item on the left wall is a bench grinder on a stand.

Along the right wall is a two-stage air compressor with an 80-gallon tank. That compressor feeds two reels holding over 50 feet of air hose each. One reel is near the side door; the other is by the rear roll-up door. Next to the compressor are a bulk engine oil tank with a hand pump, oxyacetylene cart, stainless steel parts storage, a workbench cabinet, and battery charger.

One of the truck's best features, Marvin feels, is its rear hydraulically operating liftgate. It not only provides easy access to the body, but also acts as a work surface. ∎

The generator set supplies power to the entire truck, which is wired as a shop would be. Tools are securely fastened to the wooden floor to prevent movement during transport.

"Wooden slats reinforcing the body's walls provide a great place to hang various items like tow ropes, cords, and hoses," Marvin Mechtel points out. "They also made it easier to wire the truck."

One of the nicest features of the Mechtel truck is a Dur-Tek stainless steel combination cabinet and workbench complete with a hood that folds down to contain items on the work top.

Service Truck for Boy Scouts

Key welding equipment – including separate 250-amp stick and wire welders – are accessible on a tailgate platform that lowers to the ground. Positioned ahead of the welder is an EdwardsJaw IV Ironworker that handles all metal preparation chores in the field. Ample power cords reach hundreds of feet.

Ultimate welding-center-on-wheels allows in-field repair and fabrication

By Dave Mowitz, Machinery Director

Brian Esser adopted the Boy Scout code, "Be Prepared," when building his welding service truck. The Minnesota Lake, Minnesotan, had plenty of experience in that regard. This is the sixth service vehicle he has created.

"I've put them on trailers and smaller trucks but never a vehicle that would allow me to handle any welding chore," Esser explains. "This time I started with a 1981 Chevy C70, 2-ton truck and extended its frame out to 30 feet."

On that base, he custom-built a utility bed offering a shopful of tool cabinets all accessible from the ground. Beyond holding a complete welding setup that rivals most shops, Esser's truck will hold a small hardware store. "I keep most common sizes of screws, bolts, nuts, rod, angle iron, flat steel, and the like on hand since many times it's a long drive to a hardware store," he says.

Tucked in one cabinet is a 220-volt air compressor offering 150 feet of hose. The truck's generator also comes with 200 feet of cord. ∎

Everything needed for full-scale welding and metal fabrication chores is strategically packed in Brian Esser's service truck.

Photographs: Ron Van Zee

A telephone utility bed mounted over a PTO generator trailer makes a shop on wheels. Jack Morris can wheel it to any location with a tractor, which also drives the PTO generator inside the center storage area.

Trailering a Shop to Work

Salvaged parts and ingenuity combine to create a low-cost mobile shop

By Le Spearman

An old telephone truck bed provided the inspiration for a trailer that Jack Morris has found indispensable. A tribute to the Delaware, Oklahoma, farmer's ability to recycle, his shop-on-wheels provides a perfect solution for transporting welding equipment to the field.

"I had an old PTO generator on a two-wheel trailer, but I needed something that was closed in," Morris remembers. "I had a guy working for me that had this old telephone truck bed he wasn't using," says Morris. "I traded cutting trees for the bed. The two-wheel generator trailer provides the running gear."

He began work on the bed shop unit by installing the PTO generator near the center and front of the bed. This allowed Morris to connect the generator to a tractor's PTO shaft.

TRACTOR HITCH FOR FIELDWORK

Next, he modified the trailer's tongue into a bulldog hitch, which allows Morris to hook the shop-on-wheels to a tractor. "I just go wherever I want, and when I get there, kick the generator in gear," says Morris. "I can let the tractor set there and run about 1,700 rpm and weld all day."

The telephone truck bed measures 7 feet long and is 6½ feet wide. The original utility boxes on each of the bed are 18 inches high. Morris used 8-inch flat steel to build 16-inch-high and 12-inch-deep storage units, which he placed on top of the old boxes. "I went around the seams with ½-inch square tubing that makes it look more finished," Morris explains.

The finishing touch came after Morris constructed a roof between the utility boxes by adding double doors across the back of the trailer. "Just below the doors and running across the back is a tailgate that drops down just like a pickup," he adds.

Adding the roof and doors created a weather-tight storage area that houses an old water heater recycled into a water storage tank, water and air hoses, and the generator. The generator provides power for both 220- and 110-volt outlets, which enables Morris to weld and use small tools. "In case of a power outage, I just unplug the welder, plug in the cord, run it into the barn, and I can work my milking machine," Morris explains.

HOMEMADE AIR TANK

Again, relying on his innovative recycling, Morris fashioned a length of 14-inch-diameter pipe into an air tank by cutting and welding circles of steel to each end of the pipe. He mounted this homemade tank underneath the bottom back of the trailer. An 8-inch sheet of flat iron was placed in front of the tank to conceal and protect it.

Compressed air from the tank passes through a regulator and powers air tools. In addition, the air supply pressurizes the water tank. This creates a power washer of sorts, which Morris utilizes to clean up in the field or for fire protection in case crop residue catches fire during welding.

Each side of the mobile shop is organized to hold additional equipment and supplies. Morris built two welding cubicles into the front of the trailer on each side of the center storage unit. One compartment stores an acetylene welder used for braising and cutting. The compartment on the opposite side contains a Lincoln 180-amp electric welder. ∎

Service Truck Does it All

left to right: Jim Leonard and his sons, Duane and Dean, created a service truck to suit their needs by extending the existing cab, fabricating oversize compartments behind that cab, and adding a heavy-duty bumper with twin hitches and outriggers for their crane.

Award-winning truck created from salvage materials is an original
By Dave Mowitz, Machinery Director

The Leonards couldn't find the service truck they wanted, so they built the truck they needed. The Morgantown, Indiana, farmers began their quest to create a service truck tailormade for their operation by purchasing a used 1992 Model FL70 Freightliner. A winter spent modifying that truck and equipping it with the tools and materials needed for a true shop-on-wheels resulted in a truck that is not only crucial to their operation, but also earned First Place in the Best Service Truck category of *Successful Farming* magazine's Top Shops® Contest.

TRUCK ALTERATIONS
The truck the Leonards chose to modify was well suited for their purpos-

es, providing a 30,000-pound gross vehicle weight rating and 22-foot box bed. But the van body that came with the truck didn't figure into their plans. "So off came the van body," explains Jim Leonard, who farms with sons Dean and Duane.

"Since we didn't need all that much frame, we decided to shorten it by removing 9 feet from the center of the frame to shorten the truck's wheelbase. In doing so we had to remove a section of the driveshaft. We also removed 5 feet from the rear of the frame (behind the rear axle)."

Even with those modifications, the truck still didn't meet all their needs as a transport vehicle. "We had seen a similar Freightliner that had a double cab," Jim recalls. "We really wanted that extra room but couldn't find a similar cab. So we found another cab identical to the unit on our truck, which had been burned and salvaged."

The Leonards took to removing the back paneling and window from the truck's existing cab. They removed the dashboard panel from the salvaged cab. Then came the tricky part of marrying the two cabs together. "We were concerned that they would not marry up together that well, but we found that wasn't a problem," Duane recalls. "We had to add some sheet metal and a few supports. But on the whole, it came together well."

REMOVE TO TAKE A CREW TO WORK

After the expanded unit was painted, the front bucket seats replaced, and a bench seat mounted in the crew part of the cab, the Leonards had created a vehicle that could transport "six grown men to the field," Jim reports. "We've even used the truck for transport when going to farm shows."

After finishing the cab, the Leonards set about mount-ing a utility body on the frame. "The used body we had came up short, lengthways, on the frame," Dean explains. "To fill that gap, we built two tall service cabinets positioned directly behind the cab."

BODY REINFORCED TO HANDLE A CRANE

Another embellishment to the utility body was the addition of a 5,000-pound-capacity IMT crane. The body wasn't constructed to handle a crane, however. So the Leonards reinforced the right rear corner of the body with a 2×4-inch tube steel boxed structure mounted to the truck frame. The top of that structure, which resides inside the right rear storage compartment, was topped off with ¾-inch plate.

The addition of the crane required outrigger wheels for stability. Those outriggers were fabricated from 2×4-inch tubing that slides into a homemade bumper. The vertical part of the bumper is fabricated using ¾-inch plate upon which a pintle hitch is mounted. The horizontal step is built of ½-inch plate and is home to a DMI spring-loaded drawbar hitch.

Other features on the Leonards' truck include a 225-gallon diesel fuel tank, hydraulically operating 110-volt generator, and a welder generator. ∎

If you didn't know any better, you would never have guessed that the crew cab on the Leonards' service truck was fashioned from their truck's existing cab expanded with the addition of a second cab on the rear. A bench back seat provides room to transport four men in the crew quarters.

The oversize cabinets behind the cab were built by the Leonards to fill the space on the frame behind the cab and ahead of the utility body. The cabinet on the driver's side stores welding equipment, air tools and their attachments, heavy-duty bolts and nuts (in sizes most used on the farm), welding rods, tools, and helmets. The passenger-side cabinet holds air hose from the compressor, the oxygen-acetylene torch and its hose (which is wound up on a reel that swings out of the cabinet), an array of air attachments, and various tools.

Super Press

Nick (left) and Randy Dolezal created the ultimate press from scratch.

This unique machine has two rams that travel up to 8 feet

By Dave Mowitz, Machinery Director

U ntil the Dolezals thought outside the box, one hydraulic press was, more or less, just like the other in terms of basic design. Randy Dolezal and his son, Nick, however, wanted more – more room, more flexibility, and more oomph when it came to a press.

The Danube, Minnesota, farm team envisioned a press with two hydraulic rams (not necessarily unique) and a design that allows those two rams to be easily positioned anywhere along the press's table (that's unique).

"Not only do the two cylinders accommodate different jobs, but we can also use the secondary cylinder to hold our work down while pressing with the other ram," Nick says. "This feature is a lot handier, and I feel safer working around the press."

SLOWLY EVOLVED

The Dolezal press was 20 years in the making in Randy's mind. "And then it took about 10 years to collect all the parts I needed for it," he says.

"Plus, I spent time on and off for the next five years building it."

That time spent was well worth the investment. It resulted in an impressive one-of-a-kind hydraulic press. So it's not surprising that the Dolezals' hydraulic press was selected the Best Shop Feature in the Top Shops® Contest.

MYRIAD OF UNIQUE FEATURES

Features don't end with the movable dual cylinder arrangement on the Dolezals' press. The tool's frame is extra wide, with the table measuring 8 feet in length. Plus, the press was built to allow for 50 inches of working width between the legs.

Furthermore, that table extends beyond the frame 32 inches at one end of the press. This offers a very handy table extension with 180° of working access.

Note that the Dolezals created a 24-inch-wide table. This extra room offers more working space and also provides greater stability of work pieces during presses.

The anatomy of the press consists of a table and header fabricated from 8-foot lengths of 12-inch I-beams with ½-inch wall thickness. Two pairs of I-beams are spaced 24 inches apart and then boxed at each end to both the table and header.

The cylinders in use consist of a 5×40-inch unit with 49-ton capacity and an 8×50-inch ram with 125-ton capacity. Both cylinders are rated at 5,000 psi).

"That is a lot of press capacity, and I was concerned whether the bare I-beams in use in the table and header could handle the load, even doubled up as they are," Randy observes. "So we reinforced the side of the beams with 12-inch-wide steel plate ½ inch thick. It took extra time welding all that plate on, but it was worth it. We don't want to bend the

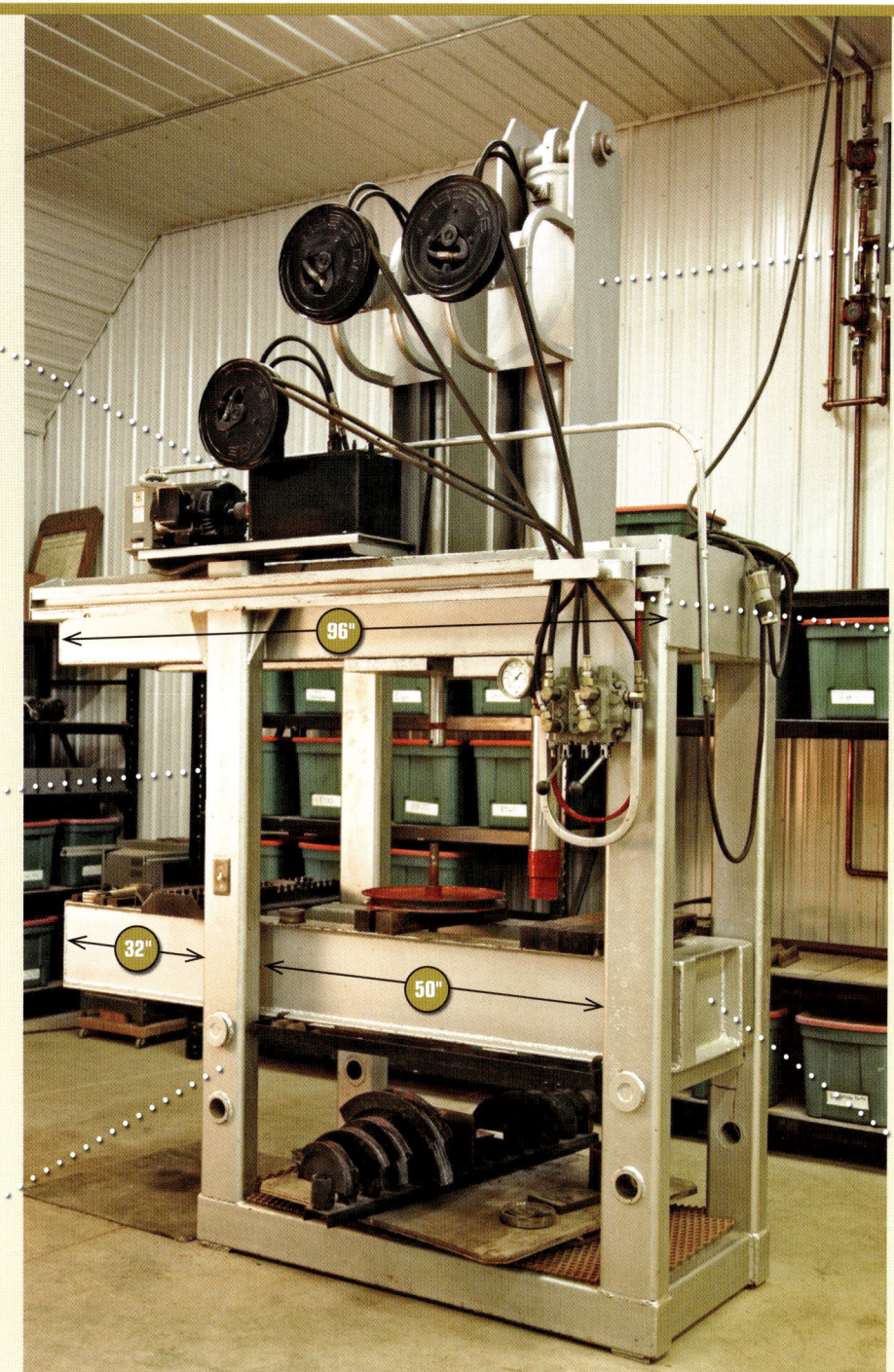

Pressurizing the press is a William eight-piston, two-stage pump powered by a 5-hp. motor. Fluid is transferred through hose reels salvaged from old forklifts. Two reels feed the cylinders, while a third feeds the controls.

The legs are formed from 7×4-inch tube steel that is ⅜-inch thick.

Table height can be adjusted to two positions utilizing pins made from 2½-inch shafts. A third work position uses the frame's base.

The cylinders are nestled inside plate steel housings mounted on massive plates. These plates ride on 3-inch-diameter rollers on a track on the bottom of the header.

Controls travel the 8-foot length of the table on rollers riding inside a track. A pressure gauge provides an instant read of cylinder capacity.

The table and header are formed from 12-inch-tall I-beams. The sides of the beams are reinforced with ½-inch plate for strength.

96"

32"

50"

press because it's pretty much scrap iron after that."

CYLINDERS ROLL TO WORK

What is ingenious about the Dolezal press are the rolling rams.

The two cylinders are nestled inside plate steel housings welded to massive plates. These plates ride on steel rollers that follow a track inside the header. In addition, the controls ride along the length of the press on a

track attached to the header.

"This allows us to position the controls near the work to start with," Nick says. "Then we can back off while pressing for safety's sake."

The Dolezals have found that they

The pump system employed by the Dolezals to pressurize the press can deliver as much as 5,000 psi. "We looked at pumps used on commercial presses, but they didn't deliver enough pressure. I've never had the pumping above 3,500 psi on all the work we've done," Randy Dolezal points out. "Also, we used a two-stage pump that runs at a higher flow rate up to the point that we contact the work. Up until then, it runs somewhere around 500 psi. When the second stage kicks, the cylinders slow down but deliver more pressure."

employ the smaller 5×40-inch ram to handle most of their repair and fabrication chores since "it is faster and can roll out to the very end of the extension of the press. We typically use it for tubing or flat metal for fabrication," Randy says.

"Then, too, the smaller ram takes care of removing most bearings, pulleys, or sprockets from shafts. But the big ram does come in handy when working on heavy material," he explains.

INVESTMENT HAS ALREADY PAID OFF

Randy calculates that he spent roughly $4,000 building the press. He employed salvage metal and parts when he could find them.

But the press has already covered that expense and then some.

"Last spring we had two tractors needing rear axle bearings replaced," Randy recalls. "The dealer wanted $3,500 each to repair them. We did it ourselves, spending less than $1,000 for parts."

Nick is still surprised at how frequently they use the press. "Almost every day we seem to use it for pressing out a bearing, making a corner on square tube, or bending round stock," he says. "It's tough to remember all the projects we've done with the press." ∎

The Dolezals also fabricated a wide assortment of bits to fit both square tubing as well as round stock. The bits come in a variety of widths to accommodate various common sizes of metal stock.

Shop Bowl

Bowling alley lane finds a second life as an Ohio farmer's workbench

By Denny Eilers

Y ou could toss a 16-pound bowling ball on the workbench in Chris Ramsey's Oxford, Ohio, farm shop and not even make a dent. That's because the bench was once in a lane in a bowling alley. "They were renovating the lanes at my local bowling alley," Ramsey recalls. "So I was able to get some of the flooring for my workbench."

The resulting bench – measuring 4 feet wide by 37 feet long – is the centerpiece of a shop Ramsey spent years planning. "For years I read every article I could find about shops," he says. "That motivated me to build what I wanted based on the best ideas from that research."

The 37-foot-long workbench is topped off with an actual bowling lane. It nestles under a balcony that serves as Chris Ramsey's office. The loft also offers parts storage for small equipment.

SHOP BUILDING DETAILS

For example, to create the building's base, Ramsey says, "We dug down 10 inches and placed a plastic layer on the ground. Over that we added 4 inches of fill, then 6 inches of concrete with mesh."

Next came a 24×32-foot structure that was insulated with Tyvek home wrap to seal off the interior. Walls were insulated first with 1½-inch blue foam board placed between the structure's posts. That was topped off with 6 inches of batt insulation.

"Don't skimp on insulation," Ramsey advises. "Be sure to put enough in when you build because it pays in the long run."

His insulation investment keeps the annual heating bill at about $120 a year. Ramsey uses propane.

One end of the shop serves as the structure's repair bay surrounded by storage cabinets and the bowling-alley workbench. Beneath the workbench are 19 drawers and five storage cabinets lining the walls. A grinder table is on rollers. "I like to move it to the door to keep the mess outside when I'm doing grinding work," Ramsey says.

For another $300, he installed an alarm system that triggers a warning siren. "There are a lot of expensive tools in a shop," he says, "and this is a good way to protect the investment."

DO-OVER LIST

Like other farmers who've built new shops, Ramsey has a do-over list that includes adding a bathroom, installing a heated floor, and putting in more electrical plugs.

"Start with a good plan," before building a shop, he advises. "Make a diagram and move things around until you get it just the way you want, especially the wiring." ∎

Convertible Doors

Creative engineering results in two doors in one opening

By John Dietz

The 26-foot-wide opening in the Butuk shop is filled by two doors. The 10-foot-wide door on the left is often used in the winter to reduce heat loss. The post between the two doors is movable, allowing wider equipment to enter.

Building a shop is often routine. Pick what you need, hire a contractor, pay the bill. For Fred Butuk, that's not quite the case.

Butuk was an engineering technologist between 1978 and 2005. His name is on about a dozen patents for air seeding and no-till systems.

In 1990, Butuk and his brother, John, picked out a shop and hired a contractor to build it on their farm near Insinger, Saskatchewan. They chose an all-steel building with a 26-foot-wide opening for a future door.

Filling that 26-foot hole became an issue for the brothers. Opening it in the winter would waste huge amounts of heat.

TWO DOORS IN ONE OPENING

So Fred Butuk put his inventor's hat on and came up with a novel solution. Instead of installing one large overhead door in the 26-foot space, he innovated a movable post that allows the use of two doors.

One door is 10 feet wide while the other is 16 feet. The post dividing the two doors can be readily moved to the side after both doors are open. Doing so provides a full 26-foot-wide opening. Or "we can open just the 10-foot-wide door if we want to bring just the pickup in during the winter," he explains.

The movable post was made of 4×8-inch rectangular tubing. The bottom of that post is held in place with a 2-inch-diameter pin (see picture on the next page). That pin is pushed down into an anchor in the concrete drive using an air brake pot from a truck. Applying air to the pot causes the pin

The doors are raised using a counterweight system.

to raise and free the bottom of the post.

The top of the post will eventually ride on a trolley rail. It is currently being held in place with bolts. But Butuk is planning on using another air brake pot to raise the post after it has been freed up at its bottom. Once raised, the post will then be able to be pushed to the side while riding on the rail.

The doors that operate on either side of the post were custom built by the Butuks. They hired a steel supplier to cut the sheet metal (20-gauge galvanized steel for exterior panels) in 10- and 16-foot lengths. With a press, he bent the outside face panels to make four sides. He mounted spars inside these panels to keep them rigid.

HEAVILY INSULATED PANELS

Two layers of Styrofoam insulation are inside the exterior panel. Interior-facing panels are riveted to the exterior steel. The resulting doors, which are 14 feet high and 4 inches thick, are heavy duty, with the emphasis on heavy. For example, the 16-foot-wide door weighs as much as a commercially built 24-foot-wide door, or around 1,350 pounds. The 10-foot door weighs about 840 pounds.

Normal wound-up door springs weren't powerful enough to assist in lifting the door. And keeping the doors up and preventing them from crashing to the floor was a concern.

So Butuk fashioned a counterweight system to help lift the doors. He poured concrete into 2-foot-square steel frames to create the weights. Each panel in the doors required one such weight.

Each counterweight for the 16-foot-wide door was poured with concrete 6 inches deep. The weights for the 10-foot door received 4 inches of concrete. The resulting weight was just right to counterbalance the weight of each respective panel.

Spacing the counterweights in a steel tube truss required careful thought and basic math. Butuk determined each counterweight had to travel 17 inches to lift one panel by 24 inches. In addition, the lifting force had to transfer horizontally to the door and then downward to the sides of the door panels.

HAND CRANKS RAISE DOORS

"A lot of thought went into the cable system to get that motion to the doors," he explains. "The counterweights don't travel as much as the door panels, so each one has to exert more force on the cable than the weight of the panel itself."

A sprocket and chain drive serve as a temporary mechanism to raise the counterweights in the truss stand. "The hand crank turned out to work quite good," Butuk reports. "But I will convert the cranks to electric motors in the future." ∎

Hand cranks will be replaced with electric motors in the future. Fred Butuk says it takes little effort, however, to raise the doors.

The bottom of the center movable post is held in place with a 2-inch-diameter pin.

Long on Practicality

Tom (below) and Doug Burrer took a no-nonsense approach to equipping their shop. A salvaged hydraulic lift works with floor cherry pickers to substitute for more expensive overhead hoists.

The Burrers build rather than buy their shop needs

By Dave Mowitz, Machinery Editor

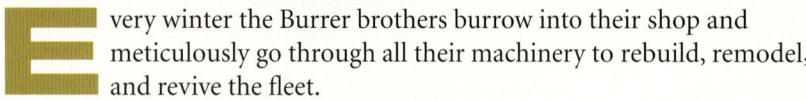

Every winter the Burrer brothers burrow into their shop and meticulously go through all their machinery to rebuild, remodel, and revive the fleet.

That dedication to maintenance explains how the Elyria, Ohio, farmers can operate with equipment that is often 20 to 25 years old. "We take good care of it," Tom Burrer explains, "and it keeps operating beyond traditional trade-in life."

EVERY INCH PUT TO WORK

Tom and his brother, Doug, credit their shop for making such maintenance diligence possible. That structure is the epitome of practicality. Every square inch of the shop is put to use whether in service bays, supply and tool storage, or a fabrication center.

Their approach to creating the shop? "We build it ourselves or salvage what we can before buying shop accessories like storage areas," Tom says. "Yet the shop has all the tools, equipment, and adequate room to perform any and all – I mean *all* and not just some – maintenance jobs."

For example, the Burrers didn't invest in an overhead swinging or trolley hoist. Instead, they purchased a salvaged hydraulic automotive floor hoist. "We have used it to lift the front of semitrucks as well as entire cars or pickups," Tom says. "Cherry picker hoists lift everything else. Sure, an overhead hoist would be great, but we couldn't justify the cost."

Such is also the case with wanting a larger structure. "You always want a larger shop, right?" Tom asks.

Yet the 48×62-foot structure can house up to three pieces of machinery

at a time for servicing. That's possible because of the way the brothers allocate floor space.

"We created separate work areas that don't interfere with each other for doing a variety of maintenance and service projects," Tom says.

THREE BAYS AVAILABLE

Two major service bays are located behind the shop's two overhead doors. A third bay is carved out around the hydraulic floor lift. "We can be changing the oil on a pickup on the hoist, for example," Tom explains, "while doing an engine or transmission overhaul on a tractor in another bay. And dirty jobs that need washing are positioned in the bay at the end of the building and near our steam cleaner."

The Burrers prefer to wash their machinery outdoors. But an indoor wash bay, equipped with a floor drain, is a necessity in the winter.

Photographs: Andy Sacks

Also, this same location is used when any painting needs to be done.

This division of work space allows the Burrers to call in a professional mechanic to tackle engine overhauls when necessary. "This certainly cuts down on the cost of an overhaul," Tom explains. "We can do one here for little more than the cost of the overhaul kit ($2,500 to $3,000), as opposed to having a job done at a mechanic's shop, which can run over $8,000."

MORE WAYS TO CUT COSTS

The brothers also trimmed construction expenses when the shop was erected a decade ago. They built on the foundation of a previous building that had burned down.

"There was no labor and not much help from any contractors," Doug says. "We put all the nuts and bolts together for everything, including electrical, plumbing, ventilation, lighting, hydraulics, loft storage, drainage, concrete, and welding." ∎

A low-cost battery charger wired to 15 leads provides a trickle charge for batteries to keep them in condition during storage. The batteries sit on wood shelves to prevent discharge, a common occurrence when batteries are stored on concrete.

Used oil is captured and held for proper disposal in homemade containers. Walls above and below the storage loft are lined with wood shelves and bins to keep parts in order and easily accessible.

A wide assortment of funnels, filter wrenches, and coolant equipment drips excess fluid into a trough made from a length of roof gutter. It resides next to the jugs of specialized oil in the Burrer brothers' lubrication center.

A centrifugal fan, salvaged from a forced-air furnace, readily sucks welding and grinding smoke then propels it down a 12-inch-diameter duct to the outdoors.

Even drain tile has a second purpose. Here it's a home for long-handle tools. Pry bars, tire irons, and sledge hammers are organized in the tile containers.

The loft serves as home for 440 gallons of lubricants in the form of engine and hydraulic oil. Two sets of drums are dedicated to a specific lubricant, which is fed to a central location on the ground floor to be tapped for use.

Precision Lube Center

David Mitchell designated a 15×30-foot area in the corner of his shop as a service center, which features a lubrication storage and retrieval system that precisely doles out fluids to waiting vehicles. The bay is also equipped with an auto hoist, overhead bulk storage, engine exhaust ventilation system, and epoxy-sealed floor.

Innovative engineering and an eye for detail create the ultimate lubricant center

By Dave Mowitz, Machinery Editor

When it comes to maintenance chores, David Mitchell wants it low maintenance. "I do not hire any outside help and do all of my own service and repair work," the Houghton, South Dakota, farmer explains. "I need to work as efficiently as possible to get everything done."

That's why Mitchell took extra care designing and equipping a service bay in his 60×104-foot shop. And his efforts paid off when he earned top honors in the Best Lubrication Center category of *Successful Farming* magazine's Top Shops® Contest sponsored by Lincoln Industrial.

The heart of Mitchell's service bay, and certainly one of the most unique features you will find in any farm shop in the country, is a bulk oil dispensing system that employs two 7½-gallon volumetric tanks that Mitchell purchased from a farm chemical dealer.

Engine (15-W40 diesel) and hydraulic oil are held in the individual 140-gallon bulk containers mounted in a storage loft. Each bulk container

At the heart of David Mitchell's service bay are two volumetric measurement tanks that receive lubrication of overhead engine and hydraulic oil bulk storage containers.

is plumbed to its own volumetric tank via 1-inch-diameter, clear and braided hose. Oil flows from the respective bulk container through the hose and past a 1-inch valve that is part of a three-way valve junction Mitchell located at the bottom of each volumetric tank (see page 40).

After oil is measured in the respective volumetric tank, Mitchell closes the fill valve from the bulk container. To release the measured oil, he has the option of opening one of two valves in the three-way junction. One valve feeds a 1-inch braided hose that Mitchell unwinds to reach waiting vehicles.

Or, he opens a third valve that drops oil down into a storage cabinet and into a waiting funneled fill can. "I use this method when doling out smaller amounts of oil like for a small gas engine," he says. "Otherwise, when servicing trucks or tractors, I almost always use the hose."

USES AIR TO PROPEL LUBRICANT

To assist in that effort, Mitchell plumbed air lines to the top of each volumetric tank. He can then charge the respective tank with air to push lubricant from the tank and through the hose. "I put a regulator at the air control valve to keep pressure down to 10 psi. Experience has taught me that more pressure than that could cause things to blow apart."

Beside the two large volumetric tanks, Mitchell has an additional handheld measuring tank he uses to dole out smaller volumes of fluids. "This tank (which hangs on a bracket on the wall between the two larger tanks) has its own diaphragm pump to suck fluid into the tank for measuring or to propel it out," Mitchell adds.

Besides measuring fluids precisely, Mitchell's engineering helps keep the service bay clean. "It makes the job of measuring out and transferring oil

Lubricant measured in the volumetric tanks can be transferred through a 1-inch hose (coiled up by David Mitchell's right hand) to waiting vehicles in the auto hoist or tractors parked next to the hoist. Oil is pushed out of the respective tank and through the delivery hose using air pressure. Or, Mitchell can drop lubricant through a metal storage cabinet and into waiting funneled oil cans.

This close-up shows the plumbing junction David Mitchell devised. It allows him to fill the volumetric tanks from bulk containers by turning the lever on the right. Measured fluid is delivered through a hose (left lever) to vehicles or dumped into quart oil cans (bottom lever).

An old propane tank is parked just outside the door that is located next to the service bay. It was converted with an oversize funnel to securely hold waste oil.

fast. And it's completely contained, which greatly reduces spills."

Additional specialized lubricants, fluids, and supplies for the service bay sit on shelves on the wall next to the measuring tanks. "I used shelves to store these items vs. cabinets with doors so I can glance over and quickly take inventory of supplies," Mitchell explains. ∎

Portable Fluids

Three 65-gallon polyethylene containers are mounted on a structural steel frame. Each tank is plumbed into Lincoln Model 4475 pumps that operate off the shop's air system. They deliver fluid through individual 25-foot-long hose reels to metered nozzles that dole out exactly the amount of fluid programmed into the touch pad. "The huge advantage to the portable lube center is that I don't have to move vehicles to fluids but can leave them in whichever bay in the shop they were parked in," says Randy Miiller.

Mobile fluids center rolls 195 gallons of lubricant and antifreeze next to vehicles

By Dave Mowitz, Machinery Director

Randy Miiller brought the mountain to Mohammed, so to speak, when it came to lubricant storage. Rather than mount his lubricant containers on permanent racks located in the corner of his shop as is most commonly done, the Mount Vernon, South Dakotan chose to make his fluids mobile.

His three 65-gallon containers sit atop a steel frame that rides on 6-inch, heavy-duty castering wheels. "The entire unit easily rolls right up to where we are working on a vehicle," Miiller points out.

UNCONVENTIONAL APPROACH

Miiller's out-of-the-box thinking about lubrication storage and retrieval not only created great convenience, but also earned him First Place honors in *Successful Farming* magazine's Top Shops® Contest.

The unique lube center was concocted when Miiller constructed a new shop several years ago. "The shop was designed with multiple work bays so it would accommodate several trucks, tractors, or combines at one time," Miiller explains. "My farm crew and I debated making the lubricant storage stationary, but we saw the drawbacks to that; namely, it would tie up a portion of the shop only for oil changes."

Also, Miller explains, a portable system lets his farm's crew roll fluids to vehicles that need engines, transmissions, or radiators topped off.

"At first, the system only had containers for engine and hydraulic oil. Then I added a third tank for antifreeze," Miiller says. "I'm surprised how much we use the antifreeze now. It sure comes in handy when we're topping off a tractor or combine with oil. Most vehicles seem to need a couple of squirts of antifreeze."

The portability of Miiller's innovation certainly is worthy of contest recognition. But it was the entire system – onboard pumps, hose reels, metered nozzles, as well as a unique waste oil collection system – that earned Miiller top contest honors.

Each of the lube center's three polyethylene tanks feed their respective fluid to individual Lincoln Industrial Model 4475 pumps. These pumps operate off Miiller's shop air. So each pump is equipped with an air regulator and air filter.

"We just attach the female end of an air hose to quick couplers that feed the individual pumps," Miiller explains. "This way the operator is forced to choose the pump needed to access a specific fluid to avoid a mistake in choosing the wrong lubricant," he adds.

Fluid is pumped from the tanks to reels each holding 25 feet of hose. The reels are bolted to the caddy frame just in case they need to be removed for maintenance.

Fluids eventually reach electronic lube meters located at the other end of each hose. Although expensive, the meters have paid for themselves "many times over just for the convenience they offer," Miiller says. "We can program the exact amount of fluid needed for a refill. When it reaches that level, the meter shuts off the pump. And the meters are equipped with nondrop nozzles, which reduces spills."

The caddy is also plumbed so that fluids from the individual tanks can be gravity-fed to ball valves equipped with spigots. These outlets are used to fill individual containers that rest on pans equipped with an expanded metal grate to catch drips.

WASTE OIL DESIGN

Another unique innovation designed by Miiller that any shop could readily use is his used oil evacuation system. The heart of the idea is a simple plastic wheelbarrow that rolls under tractors to catch used fluids. An expanded metal grate is bolted to the sides of the wheelbarrow to create a platform on which oil filters can be placed after being removed from an engine or transmission.

Miiller drilled a hole at the bottom of the back side of the wheelbarrow. In that opening he mounted a ball valve equipped with a quick coupler. Once all fluid is collected from a vehicle, the wheelbarrow can be moved over to a hand cart on which an 8-gpm Tuthill transfer pump is mounted.

The wheelbarrow can then be coupled up to the pump, which sucks the used oil from the waste barrel propelling it to a waste oil storage tank located outside the shop.

"This system has done more to reduce spills and keep the shop floor clean. Changing oil is really mess-free as a result," Miiller says. ∎

Waste oil is caught by a wheelbarrow. A ball valve with quick coupler is plumbed into a hole cut into the back of the cart.

Used oil is drawn out of the bottom of the wheelbarrow with an 8-gpm transfer pump and transferred to a waste barrel.

All the air line valves, electronic meters, and spigots are labeled to differentiate between fluids to avoid mistakes. Note the rack mounted above the air pumps. Five holes were drilled into the rack to hold funnels after use. The drain pan is equipped with a valve and quick coupler to remove fluids.

Top Shops® Feature Champ

"Because of the limited space in my shop, I had to custom build this tower crane so it could be fully swung out of the way when not in use," explains Roger Johnson of his mast crane.

Farmyard full of innovative engineering goes into a combo hoist and elevator

By Dave Mowitz, Machinery Editor

Roger Johnson is a model of self-sufficiency. A farmyard of homebuilt machinery has exited the doors of his Chandler, Minnesota, shop.

Yet, the most impressive of all his fabrications is a hoist that assisted in building much of that machinery. "Because of the limited amount of shop space, I decided to custom build a crane hoist that didn't take up much room," Johnson explains.

In addition to tending to lifting duties, Johnson's hoist does double duty as an elevator to provide easy access to his storage loft.

The feature-rich design of the hoist won Johnson the Best Shop Feature category of *Successful Farming* magazine's Top Shops® Contest.

Johnson's mast of jib hoist is as homebuilt as an invention can get. The structure is fabricated almost entirely from salvage materials. For example, the hoist's tower is a 6×8-inch toolbar from a six-row cultivator. "The bottom of that bar is bolted to the cement floor. Plus, it is bolted to the framework of the loft (about halfway up the tower), while its top is anchored to the shop's structural ridge."

For the hoist's mast, Johnson fashioned an Extanda boom from two toolbars, one of which fits inside the other. The exterior bar came off a four-row cultivator and is reinforced by an overhead rod steel bridge support.

The interior toolbar came from a disk wing. It slides out 29 inches pushed by a two-way hydraulic cylinder mounted inside the exterior bar, thus earning its Extanda name.

MOVES UP AND DOWN, SIDE TO SIDE

The entire mast is raised or lowered as much as 8 feet (at the end of the mast) with a 4-inch-diameter cylinder. Plus, Johnson added a 2,000-pound

Roger Johnson added an elevating platform to his hoist to access a storage loft.

winch; its cable passes through a pulley at the end of the mast. The winch operates off a 12-volt battery connected to a trickle charger. "That winch comes in very handy for accurately lowering things very slowly into place," Johnson adds.

Finally, the mast swings right or left in a 180° arc thanks to an innovative hydraulically operated chain drive. Pressure for the cylinders as well as the hydraulic orbit motor comes from a hydraulic pump driven off a 3-hp. electric motor. Hoist control is provided through an electric-over-hydraulic valve assembly. That assembly is connected to a custom-built control box via a 25-foot cord.

As if Johnson's hoist isn't feature-rich enough, he added something that is truly unique. Running up and down the side of the mast is a man lift consisting of expanded metal and 2×2-inch tube steel platform. An 880-pound electric winch mounted at the top of the tower propels the lift up or down. For safety's sake, Johnson added a braking system consisting of "a dog-style latch that pushes against the side of the tower when the lift is going up. A pedal must be pushed in order to lower the lift," he explains.

As a final touch, Johnson mounted a hydraulic splitter valve at the bottom of the tower to deliver fluid power from the hydraulic pump to the shop floor. "This way I do not need a tractor in the shop in order to raise or lower equipment," he says. ∎

A 4-inch cylinder raises or lowers the mast. It provides the majority of the lifting power.

left: The electric cable winch (at the top) is used to raise or lower the man lift. The chain drive operates off a hydraulic orbit motor (to the left of the tower) rotating a sprocket that, in turn, rotates a shaft that swings the mast right or left.

right: The mast pivots at two hinge points on the tower: at the base of the hydraulic cylinder and where the mast meets the tower. The solid shaft from the sprocket is attached to 2-inch-square bar stock. The electric cable winch used to inch work into place is mounted on the top of the mast.

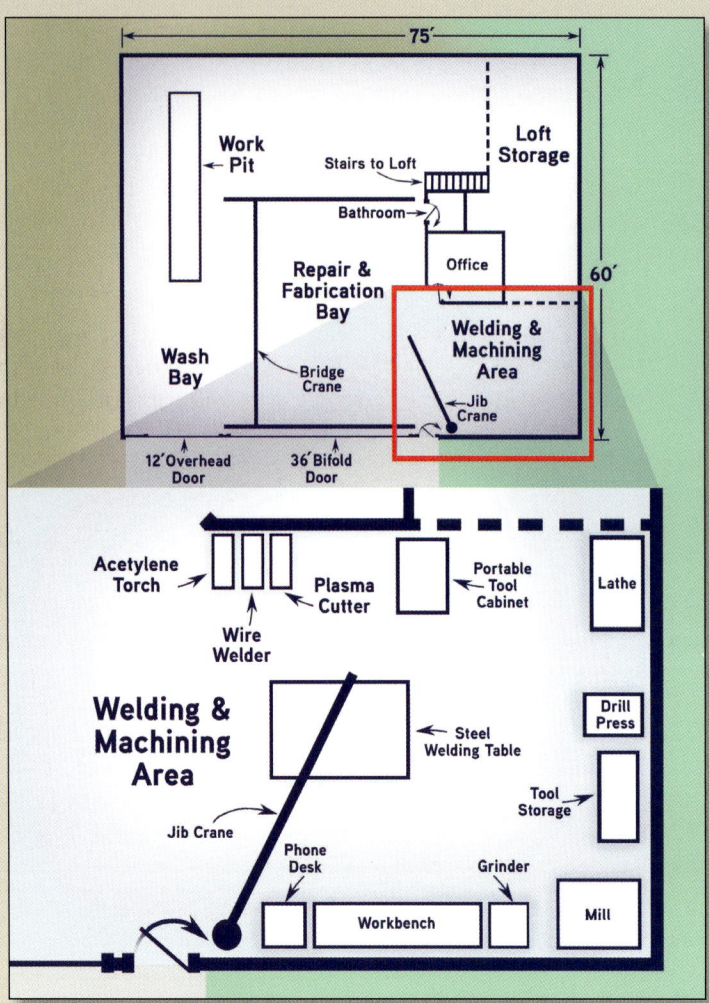

The Gerstackers' welding and machining area sits to the side of their shop's 36×30-foot repair and fabrication bay. A bridge crane is located over the repair bay area, while a jib crane services the welding area.

"You name it and we can build it," says Clark Gerstacker. "Fertilizer tenders, trailers, boxes, and spray tank brackets are a few of the items we've built."

Earl Gerstacker's original shop was at one end of a 40×90-foot building (shown behind him). "We used it mostly for repairs," he remembers.

The main entrance into the Gerstacker shop is this 36-foot-wide bifold door located in front of the building's trolley crane. "This door is ideal for big equipment," Clark says. "But it does let a lot of heat out. So we added a second 12-foot door beside it to get smaller machines and vehicles quickly in and out of the service bay behind it."

Fit to Fabricate

Separate welding and machining area design wins family first-place honors

By Dave Mowitz, Machinery Director

Inspiration for Gerstacker Farm's award-winning shop was found in their previous building. Built in 1963, that old shop was loaded with then state-of-the-art features like spacious workbenches, multiple entrances, and a trolley hoist.

But machinery outgrew that space, stirring the creative juices in Earl, Clark, and Kirk Gerstacker of Midland, Michigan.

The father and sons set out to design a new shop big enough to accommodate growing machinery. In doing so, they built one of the best welding and machining centers ever featured in this magazine.

SOMETHING OLD, SOMETHING NEW

The Gerstackers began by noting what was right with the old shop. One of its best features was a trolley crane. A similar hoist in the new shop would need to be much taller, however, to trolley over tall equipment like combines. "That's why we opted for a steel structure," says Clark. "The clear-span trusses give much higher ceiling clearance and the strength to hold a bridge crane."

Photographs: Andy Sacks   Floor Plan: Paul Bridgford

That crane is positioned behind the shop's main 36-foot-wide bifold door. The hoist beam rides on tracks extending back 30 feet into the shop from the doorway. The entire structure is designed for a 5-ton lifting capacity. "There is a 15-foot clearance under the movable beam that holds two chain falls," Earl points out. "So we can pick up and move equipment anywhere in the front bay of the shop."

As capable as that crane was at servicing the Gerstackers' main repair bay, it could not readily move items into their prize-winning machining center. "We couldn't get heavy items over to the welding table or the mill," Clark says. "Once we had to move 300-pound weights to the mill. It wasn't enjoyable maneuvering those by hand. So we designed a jib crane capable of swinging into the main repair bay where it picks up welding work, then swings back into the fabrication area."

That crane's beam operates off a post positioned to the right side of the shop's main door. The beam swings over a massive industrial-style welding table, and also reaches out to a mill and workbench. All welding equipment is positioned near the welding table along the shop's office.

STRATEGIC PLACEMENT OF TOOLS

To create the machining area, the Gerstackers strategically positioned related equipment like a grinding stand, metal mill, drill press, and metal lathe on the walls surrounding the welding table.

One of the best features of their layout is an industrial air vacuum system that "filters air and keeps it indoors rather than venting fumes and heat outdoors," Clark says. "We were fortunate to stumble upon a used Dust Hog system, which made this investment affordable."

The Gerstackers were particularly concerned about the health hazards of both welding and plasma cutting fumes (see sidebar below). "Plasma cutters, in particular, are extremely dirty," Clark says. "The beauty of this system is that it starts sucking fumes almost immediately after pushing a button. You can direct the suction inlet right by the area where you are welding."

Another innovative feature is the use of 12-foot-long expansion joints in the floor located beside the machining area. Care was taken to make sure these grooves were formed at perfect right angles where they intersect. Now they can use the grooves to hold angle iron that serves as a floor jig for fabrication.

The Gerstackers also created a metal handling and storage space on the other side of the office – away from the welding area's sparks and fumes. This area has a storage rack, brake, lathe, 50-ton press, and metal saw. "Metal can be cut there and brought into the welding area to be assembled," Clark says. ∎

LUNG CANCER RISK IN NONVENTED WELDING AND CUTTING FUMES

Clark Gerstacker's observations about the health hazards of welding and plasma cutter fumes deserve serious consideration. The National Institute for Occupational Safety and Health warns that welders have a 40% increase in the odds of developing lung cancer. Other ailments from long-term exposure to these fumes include asthma, emphysema, chronic bronchitis, and fibrosis of the lung.

Atomization of steel by a plasma cutter creates a plume of toxic fumes. And welding metals like stainless steel, cadmium- or lead-coated steel or nickel, chrome, and zinc generates fumes considerably more toxic than those encountered with mild steel.

Ventilation is key to avoid breathing fumes. Timothy Lawrence at Ohio State University says that general shop ventilation is sufficient for welding and cutting if the welding area is at least 10,000 cubic feet (if separated from the rest of the shop) and is under a ceiling at least 16 feet tall. Furthermore, you need cross ventilation if the shop is not blocked from the welding area by walls, equipment, or other barriers.

GENERAL VENTILATION NEEDS

If these requirements are not in place, then you'll need to install ventilation equipment that exhausts at least 2,000 cfm of air from the entire welding area, except where local exhaust hoods or air line respirators are in use,

Lawrence strongly advises.

Ideal ventilation is that which removes fumes before they pass by the welder's face. This can be supplied either by a professional system like that used by the Gerstackers or with a flexible hose (6 to 8 inches in diameter) equipped with a hood. Place the hose and hood 6 to 9 inches from the welding work. The fan servicing the hose should provide 550 to 750 cfm of air movement at the hood's inlet.

Another option is an air-type respirator or respirator specially designed to filter metal fumes.

Finally, even welding outdoors is no guarantee you will be free from inhaling welding fumes. Always try to stand upwind from the welding fumes if you don't have a proper respirator. ∎

Shop Storage Headquarters

The Dammans' floor plan located their lubrication and maintenance center in the left-hand corner of the shop and next to an overhead storage loft. A massive workbench and tool storage resides under that loft. The major repair and fabrication bay, which can accommodate multiple vehicles such as a tractor and combine parked side by side, is positioned in front of the shop's 16-foot-tall by 30-foot-wide overhead door.

Supersized structure acts as this farm's one-stop shop and storage center

By Dave Mowitz, Machinery Editor

The Damman brothers had grown tired of chasing down equipment stored in sheds scattered miles apart. But that was the least of their inconveniences.

Maintenance and repair chores were being handled in decades-old, cramped sheds that could barely accommodate a pickup, let alone a combine. So Duane and Dean Damman of Melbourne, Iowa, took a look ahead at their shop and storage needs, considering future growth for sons wanting to farm. Their solution was a 72-foot-wide by 202-foot-long Morton building that fills both shop and storage needs.

Their now-spacious shop occupies 60 feet at one end of this building. In the remaining 142 feet, they store machinery, seed, and chemical. "It was not only more economical to have both the shop and storage under one roof, but also it was far more convenient," Duane points out.

"Now we can grab tools from the shop and walk next door to the storage area," Dean adds. "Everything is a short walk away." ∎

Duane Damman accesses a 440-gallon bulk oil system that is part of the shop's lubrication center. Fluid is pumped from individual tanks through a 50-foot retractable hose to a metered valve. The Dammans put their hydraulic automotive lift right by the lubrication storage to create a full-feature maintenance center.

Photographs: Dave Mowitz

The pneumatic wheels easily roll Glen Wasmuth's 2,500-pound workbench right up to the work site.

Workbench on Wheels

This portable workbench handles most repair and fabrication chores

By John Deitz

A portable bench that carries everything but the kitchen sink is what Glen Wasmuth used before he had a shop. Now that he has a shop, the Battleford, Saskatchewan, farmer finds his workbench-on-wheels even more valuable.

"I found it frustrating, running back and forth to workbenches all the time. With this, everything can be rolled right next to where I need it," he explains.

To fabricate the cart, Wasmuth started with a 30×72-inch steel plate. He added shelves: ⅜-inch thick for the work top, 3/16-inch thick for the middle, ¼-inch thick for the bottom. The plates are welded to angle iron uprights. The entire frame rides on a rear axle equipped with 12-inch-diameter tires.

A castering 8-inch wheel assembly from an old combine pickup and equipped with a 6-foot tongue is attached to the front. In the shop and on smooth ground, one person can pull the fully loaded, 2,500-pound bench. "I can turn this on the same radius as the length, just spin it around," Wasmuth says.

FULL ELECTRICAL SERVICE

Wasmuth installed an electrical panel at the end of the cart. It has breakers for the 180-amp welder, a 5-hp. air compressor, a ½-inch drill press, and a 40-amp plug where he can attach a mobile MIG welder. The bench also has separate plugs, with breakers, for power hand tools. "We can have both the stick welder and the MIG welder operating at the same time," he says.

For metal fabricating, the bench carries an oxygen tank, propane tank, propane cutting torch, welding helmets, gloves, and supplies. Along the work top, he added a 12-inch chop saw, 6-inch vise, and 6-inch stationary grinder.

Helmets plus a retractable trouble light hang above the drill press, which, in turn, is perched on a shelf over the front wheels. The shelf holds sets of drill bits as well as a portable air tank.

A full-size mechanic's tool chest is at the end. The chest fits securely in an angle iron framework. Wasmuth attached tubing beside the tool chest and curved it over the top. It holds angle grinders, an old yard light that lights the bench, and a shop radio on a spring-loaded swivel.

Two sets of drawers fit inside the workbench to hold a volume of fasteners and small parts. The front set measures 4×11 inches. The back set is 4×18 inches. ■

Trouble-Free Tire Dismounting

Don Kubly begins the heavy work of dismounting a tire by separating the tire's bead from the wheel rim. After loosening a 6-inch-long section of the bead, he'll be ready to put a bead-busting tool to work.

Remove rubber with basic tools, some brawn, and this know-how

By Dave Mowitz, Machinery Editor

Removing tires from tractor rims can be an exercise similar to wrestling a 400-pound steer. "Can be" is the operable term here. Our tire guru, Don Kubly of Gempler's, says with the right tools and liquid concoctions, you can slick a tire off in short order.

The essential starting point when attempting to remove tires is to park the host tractor on a solid, level surface. "Concrete is always best," Kubly says. "Next, block the front and rear wheels, raise the tractor with an appropriate jack, and support with jack stands. Make absolutely sure it won't move forward, backward, or side to side when a tire is being changed."

For rims that have valve stems on the outside of the tire, turn the rim so that the stem is on top. Then remove the valve core to deflate the tire. Kubly also warns that tire removal becomes a bigger job that may require more than one person and larger tools when dismounting rubber from large tractors.

BUSTING THE BEAD

After the tire is deflated, drive a tire iron between the outside bead and the lip of the rim if the tire doesn't have a pry notch. This will loosen the bead from the rim lip and make it easier to install a bead-busting tool. After loosening an area about 6 inches wide, insert the buster tool. Drive it between the bead and rim flange locking the tool to the rim. Using a socket and a ½-inch drive ratchet, turn the bead buster's screw head to force the bead off the rim seat. Repeat this procedure to loosen the inside bead.

Other handy breaking tools:
- 31-inch swan-neck bead breaker
- Ken-Tool's 46-inch heavy-duty slide hammer, or their 46-inch medium-duty slide hammer
- 17-inch short-handled bead-breaking hammer
- 11¾-inch driving iron

SPOON OUT THE TUBE

Once the tire is separated from the rim, check to see whether its tube (if so used) is stuck or rusted fast to the

Photographs: John Dietz

rim. If this is the case, Kubly recommends the use of 30-inch curved tire spoons. Use one spoon to raise the bead and a second spoon to pry the tube up and loose. The curved ends of the spoons will help lift the bead over the rim flange for removal.

After both of the tire's beads are loose, lubricate the beads and bead seat area with a lubricant like Murphy's Tire & Tube Mounting Compound, Bead Seal, or Rema's Premixed Bead Butter.

To remove the tire, lock the wheel with its valve hole on top, then push the bottom of the tire inward. Insert long tire irons under the outside bead at the top and pry the bead over the rim. Be sure to keep the tire irons close together since this makes prying easier and helps prevent bead damage.

REMOVE A SECTION AT A TIME

After part of the bead is over the rim flange, hold one iron in place and remove the other to pry the next portion of bead. Don't try to pry too large a portion at a time because that makes the job harder, and it could tear the bead. "Maintain a constant grip and pressure on the irons at all times," Kubly warns. "They may spring back if your grip is released."

Remove the tube after the outside bead is completely over the rim. To do this, start at the bottom and pull the tube outward. Work your way up the sides of the tire.

Finally, finish lifting the inside bead over the rim. To do this, insert two tire irons under the bead on the side of the tire. Apply pressure on both irons and pull back on one iron to move the bead over the rim.

After the tire is removed, be sure to wire-brush all rust from the rim and clean the surface of debris. Check for cracks on bead flanges. And check the area where the rim is mounted to the center plate. Make sure there are no sharp burrs on the rim flange that could cut the tube during installation.

After cleaning, apply a thin coat of rust preventative, such as Rim Rust Preventative, to the rim's bead seat area. This grease-type coating keeps rust and scale from forming on the rim surface. It also stops the freezing or bonding between the bead and rim seat surface for easier removal of the tire the next time the job must be tackled.

To learn more about trouble-free tire dismounting, you can contact Don Kubly at www.gemplers.com or by calling 800/874-4755. ∎

Kubly employs a bead-busting jack to make the job of separating the tire from the wheel's rim seat much easier.

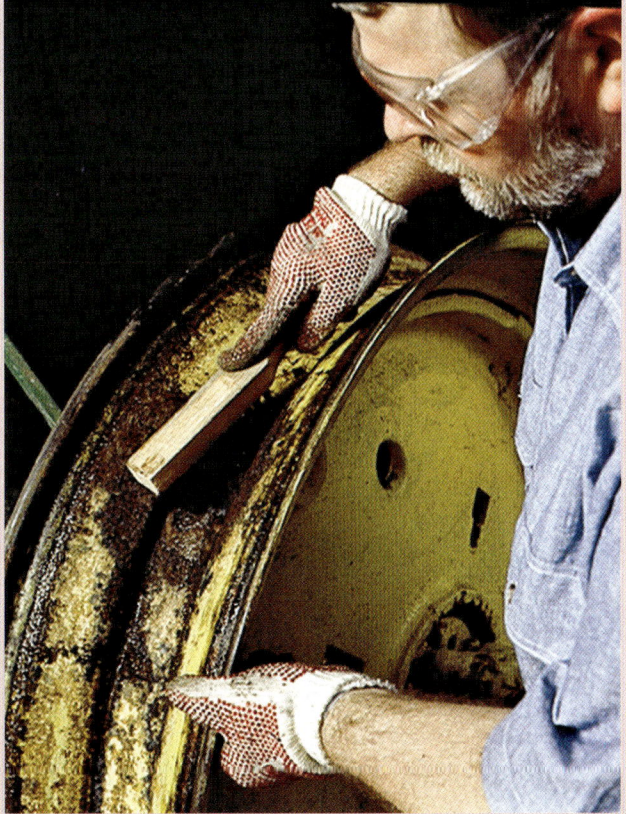

Completely clean the rim by removing all rust with a wire brush and then apply a rust-preventive.

Mounting Tractor Tires

After mounting the inside bead of the tire, insert the tube by first pulling the valve stem through the rim's valve hole. Then secure the valve stem with a rim nut, advises Don Kubly, Gempler's tire expert.

A second set of hands, a pair of tire irons, and patience can make quick work of a seemingly difficult maintenance chore

By Dave Mowitz, Machinery Editor

There are some tire repair jobs best left to professionals. The ballasted rubber of huge horsepower tractors can be too much for a novice repairperson. Still, there are many chores – like mounting small- to medium-size tubed tires – you can do with basic tools, says Don Kubly.

Gempler's tire guru and former farm-tire repairperson is a font of knowledge on the topic. And the first bit of wisdom Kubly imparts is to get help when mounting drive tires. "They can be quite cumbersome," he says. "Often these are two-person jobs, as the tires – particularly radial types – are so heavy."

Begin by lubricating the inside tire bead with a mounting lubricant. Be sure to check for arrows on the sidewall for correct rotation if the tire is directional. Make sure the valve hole is positioned at the rim's bottom.

MOUNTING THE INSIDE BEAD

To mount the tire, roll it so that the inside bead sets over the rim flange on top of the wheel. Use tire irons to slide the inside bead over the rim flange on both sides, doing the bottom last.

Once the inside bead is completely on the rim, pull the outside of the tire out at the bottom and top to make room for the tube. Pull the tube's valve stem through the valve hole and screw on the rim nut. Then, carefully slide the tube over the rim flange.

Next, put a small amount of air in the tube. Screw a valve fishing tool to the stem to keep the stem from slipping through the rim when mounting the tire. Then, remove the rim nut so the tube can move freely in the tire.

With the tube in, install the outside bead. "Again, lube the bead," says Kubly. "Starting at the top of the tire, use two tire irons to lift the outer bead of the tire up and over the top of the rim flange. Make sure the bead is not pinching the tube."

Continue sliding the bead over the rim in small sections until complete. "Trying to slide over too large a section of tire at a time may cut or damage the bead area," Kubly says.

"Next, pull on the fishing tool to bring the tube valve stem through the valve hole, and reinstall the rim nut," he says.

Before you inflate the tube, first lower the jack on the axle until the rim is centered in the tire. "This is crucial in order to seat beads properly," Kubly says. "Make sure beads and rim flanges are aligned."

SEATING THE BEADS

Using a remote-control safety inflating gauge (featuring a clip-on air chuck and air gauge measuring from 10 to 120 psi), slowly inflate the tire. Make sure the tire beads and rim flanges align properly. Continue to inflate to seat the beads, but don't inflate beyond 35 psi or manufacturer's recommendations.

When both beads are completely seated, remove the valve core and deflate the tube. Then, reinflate to manufacturer's specifications.

"Always stand off to the side of the tire when inflating," Kubly warns. "Never stand directly in front of the tire bead and sidewall area." ■

Photograph: Gempler's

Tube-Patching Basics

Patch tubes like a pro with these 10 steps

By Dave Mowitz, Machinery Editor

No more than one hour after repairing its tube did the tire on the sprayer I was using go flat. The victim of a poor patch job.

Maybe it was because I used sandpaper instead of a buffer to prepare the patch sight. Or I might have slathered on too much vulcanizing fluid. Whatever the reason, the end result was that there I sat in the field with a flat tire.

Don Kubly assures me such mistakes aren't unusual. He should know, with his 23 years of farm tire repair experience, including nine advising people on the topic of tire repair for Gempler's (www.gemplers.com). Such mistakes may be common, but they are avoidable if you know the basics of patching tubes and stick to those basics religiously.

Kubly offers these dos and don'ts for patching tubes:

- Do use repair products and instructions from the same manufacturer. "Don't use Rema patches with Camel vulcanizing fluid," Kubly advises.
- Do repair tubes in a clean, dry area.
- Don't use a cloth or paper towel to clean tubes before a patch.
- "Don't use air from a compressor to blow dust from the tube as it can contain moisture and oils, and cause a poor bond between the patch and cold vulcanizing fluids (a form of tire cement)," Kubly says.

With those warnings in mind, follow these step-by-step instructions from Kubly to tackle a patching job.

1 Inspect the tube for injury. This method of tube repair uses the Rema Tube Repair Kit.

2 Round out or buttonhole the ends of the hole using a paper hole punch or scissors to prevent future tearing as well as other rubbing damage to the tube.

far left: Avoid using buffers with speeds that exceed 5,000 rpm. **left:** Apply only a thin coat of vulcanizing fluid to the patch area. **bottom:** After pressing the patch, cover the entire area with tire talc.

3 Clean the area using a prebuff cleaner, allowing the spray to set for 10 to 15 seconds. Then drag a tire scraper over the entire surface to remove contaminants, and repeat if needed.

4 Choose a patch that extends beyond the injury by at least ½ inch.

5 Buff the repair area slightly larger than the patch size and to a smooth velvety surface. Avoid using a buffer that exceeds 5,000 rpm, Kubly warns, since this scorches the rubber surface and prevents the best bond.

6 Use a brass-bristled brush to remove buffing dust from the tube.

7 Apply a thin coat of vulcanizing fluid to the entire buffed area, making sure not to overapply the cement. Allow the application to dry, then test an area outside of the patch zone. When the cement is tacky (doesn't stick to your finger), you can apply the patch.

8 Peel foil backing from the patch and center it over the hole being careful not to touch the bonding surface. Press the patch down on the tube, starting in the center and working out to the edges.

9 Using a tire tool called a stitcher, press down firmly on the patch, once again starting in the center and working to the outer edges. This effort will remove any air that may be trapped under the patch.

10 Peel the plastic covering off the top of the patch once again starting in the center and working to the outer edges. Then cover the entire buffed and cemented area with tire talc to prevent damage to the tube. Now you're ready to put a small amount of air in the tube to check the repair before installing it back in the tire. ∎

Abrasive Cleaning Alternatives

Aluminum oxide, glass beads, or walnut shells offer better results

By Dave Mowitz, Machinery Editor

Mixtures of various abrasive cleaners (such as Jim Deardorff's Classic Blast shown left, which consists of walnut shells and aluminum oxide) not only speed up paint removal but also protect delicate sheet metal or plastic parts.

The end result of abrasive cleaning is gray metal in which delicate items like fins are not warped by heat.

As increasing numbers of farmers are taking to repainting their old equipment, Jim Deardorff has seen sales in sandblasting equipment jump in farm outlets. "Actually, the correct term is abrasive cleaning," the painting consultant says. "Abrasive cleaning is crucial preparation that can make cheap paint look great."

There are numerous abrasive cleaners on the market. These manufactured cleaners offer characteristics that are superior to sand. In some cases they are softer than sand and, therefore, less apt to erode the surface of plastic or metal, for instance.

Other cleaners have sharper edges than sand, so they can chip away paint and rust faster. Yet other cleaners are less prone to break apart upon impact and can be recycled.

"Sand's big advantage is that it's cheap," Deardorff points out. "But it creates a lot of dust, can easily pit or erode metal, and is prone to generating heat, causing pieces to warp."

The variety of alternative abrasive cleaners includes these.

- Aluminum oxide: This hard-cutting abrasive with sharp edges is very effective for removing paint.
- Crushed glass: Less abrasive than aluminum oxide, crushed glass has the hardness of glass beads but with sharper edges.
- Glass beads: The spherical shape of glass beads keeps them from cutting into metal surfaces. It is good for light deburring, putting a satin finish on a part, or removing surface particles.
- Walnut shells: Ground shells are lighter than aluminum oxide but very resistant to breaking up on impact and thus last longer.
- Plastic: The media size makes it good for stripping paint from metal without causing dimensional changes.
- Sodium bicarbonate: This is one of the softest abrasives available but requires special equipment to use.

Years of experimenting with alternatives has guided Deardorff to create a custom blend of aluminum oxide (for hardness and cutting ability) and graded walnut shells (to reduce impact breakup). That product, called Classic Blast, works well on the thin-gauge sheet metal found on cars and tractors. "It is recyclable, allows blasting at lower air pressures (which reduces temperature buildup), and doesn't erode the surface of the metal," Deardorff explains.

That last point is crucial since eroding metal encourages contamination. "You won't believe the amount of microscopic dust that sand leaves behind on metal even after it has been washed," he adds.

POWER WASH PRIOR TO BLASTING

Contamination can ruin a paint job and cause premature corrosion. For this reason, Deardorff recommends applying a degreasing product and removing it with a hot-water pressure washer at 2,500 psi prior to abrasive cleaning. This combination removes oil and grease and can make blasting 25% faster. "Blasting can remove grime, but it won't entirely remove oil," Deardorff says.

For more information, call Jim Deardorff at 660/646-6355 or visit www.classicblast.com. ■

Blast it All

Ideal sand for blasting is #00 graded that has been washed.

Don't scrimp on safety gear when cleaning metal. Stick strictly to equipment made specifically for the job, urges Dick Bockwoldt.

Master restorer shares 5 steps to clean metal with abrasive cleaners

By Ron Van Zee

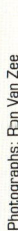
Photographs: Ron Van Zee

As a respected restorer of show tractors, Dick Bockwoldt has tried various paint removal methods like acids, liquid stripper, and rust-bonding coatings. But he has come to depend on abrasive cleaning (sandblasting) for getting equipment down to bare metal.

Bockwoldt stresses that clean metal is crucial for a long-lasting paint job. After removing all grime and oil from paint surfaces with a hot water pressure washing, he follows this five-point plan to abrasive cleaning.

1. BE SAFE

Safety is a great concern to Bockwoldt. He suits up in leather gloves and vest. Over that he wears a canvas bib front and back. His helmet has a face mask, and a separate regulator takes some of the compressor air down to about 20 psi then sends it through a replaceable cartridge filter to remove dust and carbon monoxide. The expansion of the air as it exits into the helmet provides a cool supply of breathing air.

2. PREPARE THE SURFACE

Bockwoldt protects certain surfaces prior to sandblasting, wrapping all exposed shafts and bearing seals to protect them. He cautions never to blast directly into a seal. He uses electrical clay plugs to cover holes in engines and equipment. This particular material is elastic and flexible enough to absorb the shock of flying sand without eroding away.

Frames, thick parts, and engine blocks can withstand the blast force without causing damage. But the blasting action can cause thin metal and delicate parts to heat up quickly, potentially warping sheet metal. Such blast work requires a delicate touch, greater distance from the blast nozzle, and slow going.

3. USE THE RIGHT SAND

Bockwoldt buys #00 graded sand and never reuses it since it breaks up after coming in contact with steel. Plus, used sand is contaminated with rust, grease, dirt, and grime.

4. SET THE AIR

Bockwoldt sets his compressor to deliver 120 to 135 psi at 180 cfm at the point where sand is introduced. The sand-air mixture exits through a titanium nozzle with sufficient force to remove an arm or leg, so keep people away from the blasting area.

5. FINISH THE JOB

After sandblasting, Bockwoldt blows off all metal with compressed air before bringing it into his shop (he prefers to blast outdoors). Exposed metal is immediately wiped down using a clean cotton cloth soaked with NAPA #6383 Surface Cleaning. He does wait to coat freshly blasted metal. Instead, he covers it with two to three coats of primer before applying a finish coat. ∎

Timing a Magneto

Checks and adjustment tips

By Ron Van Zee

In previous issues, magneto expert Cork Groth has taken us through the process of bringing the inner workings of this device up to snuff. With that chore completed, we're ready to first time the magneto as well as reinstall it on the tractor.

But there's one last restoration chore before timing the magneto. Cork recommends replacing the ignition switch and all electrical wiring on a tractor. In particular, replace spark plug wires with units that are made up of nonresistant copper wire and feature new brass tips soldered to the wire before reassembly.

With that job out of the way, you can set to work timing. The magneto we've chosen is a base-mount WICO. Tip this particular magneto upward and look past the edge of the impulse coupling. You'll see a V-shape trip attached to an inner adjustable index trip ring. There should be some screws holding this ring securely to the housing. This particular WICO magneto utilizes a 35° lug. Some magnetos employ a 25° lug for adjustment.

ADJUSTING TIMING

With a 35° lug, rotate the index trip ring 4½ marks left from its top, or 12 o'clock position. Next, tighten down the retaining screws. On a 25° lug magneto you need to rotate the ring 1½ marks left from the 12 o'clock position and tighten down the screws.

When Cork rebuilds a magneto, he mounts it on a test stand for a final test before shipping or installing on an engine. A magneto in good condition should produce a hot blue and white spark that extends about ¼ inch when the impulse is tripped and the magneto is turning at engine operating speed.

After this spark check, you can put the magneto back on the engine by carefully engaging the two fingers on the magneto into the appropriate coupling slots on the governor or cam shaft. One final adjustment is made by slowly turning the engine over and making sure the magneto impulse clicks at piston top dead center. Some magnetos have slotted holes on the mounting plate that provide rotation for fine-tuning adjustments.

The V-shape trip (pointed to by Cork) is attached to an inner adjustable index ring on most magneto designs. That ring is held in place with adjustment screws which, when loosened, allow the ring to be rotated for timing. Indexing marks can be found on the interior of the magneto.

Cork Groth is proprietor of the Tractor Trauma Center. His "medical services" for classic tractors include rebuilding all brands of carburetors and magnetos, as well as recreating parts. Contact him at 563/285-7009 or via e-mail at cwgroth@aol.com.

Cork is available if you choose to box up your ailing magneto and let him test and rebuild the unit. (See his contact information above.) He also stocks several parts and can build or rebuild most mechanical components. A test bench run-out also comes with a magneto rebuild. ■

The slots on the governor or cam shaft coupling must align with the two fingers protruding from the magneto when remounting on a tractor.

Photographs: Ron Van Zee

Choking Carburetor

Often the solution for a fouled-up carburetor is a thorough cleaning

By Ron Van Zee

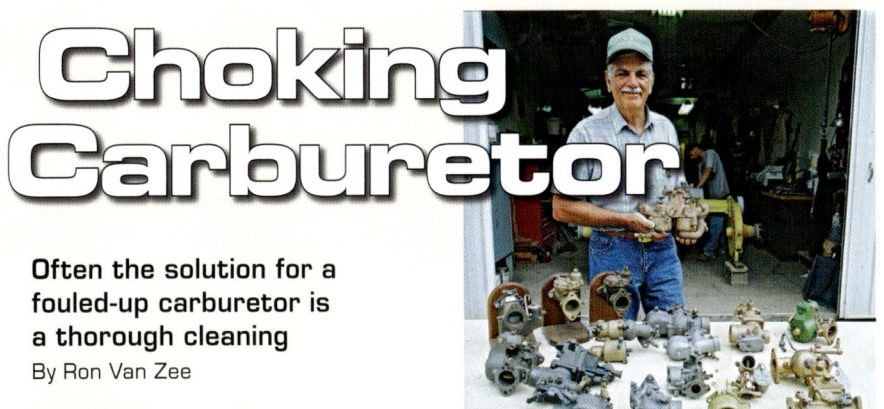

Cork Groth has made a business of repairing old carburetors such as these taken off older-model tractors. The major cause of carburetor malfunction, whether the unit is old or new, is a buildup of varnish from old gasoline. This varnish coats and clogs the many small passageways, particularly jets, in a carburetor.

Don't be embarrassed if you can't get a carburetor to click, says Cork Groth. "They are in another world of repair away from many other engine parts," the professional restorer says.

Typical dunking of metal parts into solvent to loosen corrosion or blowing out gaskets doesn't work with the fine parts of carburetors. And trying to find parts or a complete replacement carb can sometimes be out of the question depending on the age of the motor. So treat these fuel mixers tenderly, and salvage what can be saved.

What if you find yourself with a gummed-up carb? The best first step is to do a thorough cleaning. "The important thing about carb cleaning is to make sure every passage and tube is open," says Groth.

He takes carburetors – particularly if they've been sitting idle for long periods – completely apart right down to the smallest screw or butterfly wing.

Begin the process by splitting the two halves of the carburetor's body. Be gentle so that you don't damage or bend the brass float inside. Remember that there may still be a small amount of fuel inside the unit. And treat it like a fire hazard until it's completely apart and dry.

Don't throw anything away until you have the correct replacement part. "Keep notes of what came off of where," Groth urges. "It will guide you during reassembly."

Scrape off any gasket material from the bottom of the body. Now you're ready to soak the halves and parts in a carburetor cleaning solution. A gallon of such solution runs $15 to $20 at automotive parts stores. Be sure the can comes with a small parts basket (found inside), a must for cleaning small parts.

AGITATE DURING SOAKING

It helps to agitate the body and parts during soaking to move the solution through the passageways. And don't be afraid to use a small brush to work away varnish buildup.

After soaking, you'll need to clean the brass jets located inside the carburetor body. Remove the jet entirely. Because some jets are pressed into place and can't be removed, you'll need to clean them in place. Avoid using metal objects for this job.

Instead, use softer tools like nylon fishing line (mono filament) to clean out the jets. Use compressed air at the same time. Blow in the opposite direction of the fuel flow to avoid jamming goo and dirt into openings.

Another trick in final cleaning is to use aerosol brake cleaner to remove residue. The key? Get into all those hidden passages where varnish is lurking. Parts cleaner will dissolve most of that varnish; compressed air will blow it out, Groth says. ∎

ABRASIVE CLEANING IS BEST FOR CARB EXTERIORS

Bead media is a softer abrasive, nonchemical means to safely clean fragile metal on the exterior of some carburetors. This approach uses compressed air sprayers to shoot a stream of micro-abrasion beads – even wheat starch – through nozzles to remove paint, grime, and rust but leave the subsurface metal largely untouched.

Since the spraying creates dust and other airborne particulates, the abrasive cleaning is typically done inside a blasting box. The media is often continually recycled during the blasting operation.

How well paint and rust are removed from carburetors depends on the hardness of the bead media, the compression speed, and both the distance from the carburetor and the angle of approach.

Mechanics worry about subjecting a full range of engine parts to a dip in chemical solvents, so the bead media process can be useful. But Cork Groth says to be certain that all blasting media is removed from the carburetor.

Buying your own bead media system is a sizable investment, costing $1,000 or more. ∎

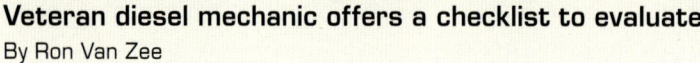

Diagnosing a Diesel

Veteran diesel mechanic offers a checklist to evaluate

By Ron Van Zee

Dave Ferguson (left) has made a life repairing engines, particularly diesel power plants. Old, dirty fuel is often the leading cause of engine problems as it wreaks havoc on injectors and fuel pumps (above) causing excessive black smoke during engine operation.

A good deal on a used tractor is not so great if you end up spending as much on that diesel as its sale price because the power plant is suffering from abuse, warns diesel guru Dave Ferguson. The 40-plus-year engine repair veteran from Austin, Minnesota, has learned a thing or two about diagnosing diesel malfunctions. His first word of advice? Always start a tractor up and listen to it operate before buying.

And it doesn't take a toolbox full of diagnostic equipment for a good evaluation. Ears and eyes, Ferguson says, are "great evaluation tools – listen for uneven compression or look for incomplete combustion, for example."

LISTEN WHILE CRANKING

A faulty cylinder weakened by excessively worn or broken rings can be detected by an uneven rhythm when turning the engine over. When a loose cylinder gets its turn at compression it will offer less resistance to the starter. This produces a loping, uneven rhythm while cranking.

Confirm a weak cylinder by cranking the engine and listening at the crankcase breather or the oil add inlet for the telltale sound of escaping air, a sign of cylinder compression loss. "The combination of escaping air and uneven cranking indicates least one weak cylinder," Ferguson adds.

If the engine has hot or glow plugs, engage them during a cold start, and listen for uneven firing upon startup. This can indicate a weak cylinder, one or more faulty glow plugs, a faulty ground wire, or bad wiring to the plug.

If the engine starts up eagerly and produces a small puff of light blue smoke, comes right to idle, and purrs like a kitten, then the chances are good that all the cylinders are generally in good working condition.

But if the engine runs rough after it's started, listen for uneven firing, and watch the color of exhaust smoke. Dirty or worn injectors can contribute to hard starting and uneven firing after cold starts. Poor cylinder compression can also cause uneven firing. Ferguson cautions, however, that old or poor fuel can induce characteristics that mimic fuel pump or injector problems.

Engine missing can go away after the engine is warmed up. If it does, Ferguson suggests you operate the engine at higher rpm and, if possible, drive the tractor to put it under load. Either action may cause the misfiring to return, which can indicate poor fuel, a dirty injector, or a weak cylinder.

With the engine running, examine the smoke it produces. Black smoke indicates unburned fuel, not a good sign at idling speed. And at higher throttle without a load, the engine should always burn clean. "A strong engine may smoke black when throttled up to high speed as well as under heavy load," Ferguson adds. "But the constant presence of heavy, black smoke indicates unburned fuel."

Also check for smoke coming from the engine's crankcase breather, the presence of which indicates cylinder compression loss due to piston ring blowby, Ferguson explains.

DETECTING TURBO WOES

If the engine is turbocharged, listen at idle and after warm-up for erratic, rattling sounds. Such noises are a sure sign the turbocharger's bearings are worn, which may call for repair or replacement.

With the engine off, walk around the tractor to look for signs of abuse, poor maintenance (old filters, for example), unorthodox repairs (jury-rigged wiring), excessive grime buildup, leaks, or dirty oil.

And note engine hours. But realize, Ferguson says, that well-cared-for diesels can have over 7,000 hours and still have a lot of life left in them. ∎

Troubleshooting
Powerless Diesels

Dirty and plugged filters are the leading culprits in stealing power

By Steven Parks

The primary causes of power loss in diesel engines are plugged fuel and air filters. So replace filters first before investing time and money in a major repair.

Diesel engine as sluggish as molasses in January? Before taking it to the dealer, spend some time with the power plant's fuel system. The cause of your problems is very likely low primary fuel pressure.

An engine starved of fuel, whether it has a computer brain or not, will cut out under heavy load, hesitate when accelerating, start hard, and warn of low fuel pressure.

All these symptoms sound ominously expensive to fix. But likely all you need to do is replace the fuel filters. In very cold weather, wax crystals will form in untreated diesel fuel, clogging filters and reducing fuel flow. When this happens, it's a struggle to get the engine thawed out and running again. Cold-weather fuel additives do a good job preventing this problem. But, to be effective, they must be mixed into the fuel before the temperature drops.

HOMEMADE FUEL LINE PURGER

Air must be removed from the fuel supply system after installing new filters. And a neat way to accomplish this is to use a 1- or 2-gallon hand-pump sprayer rigged with a fuel hose, shutoff valve, and interchangeable fittings, which connect into the primary fuel filter housing.

Using this homemade tool is simple. Pour some diesel in the tank, pump it up, hook it up to the primary filter, and open the valve. This forces fuel into the system and pushes or purges all air out. An added benefit: it's also a great way to get winterized diesel into an engine stalled by untreated fuel.

Exhausted filters figure big as the leading causes of engine problems. So while you are replacing the fuel filters, do your engine a favor and install a new air filter. Too much black smoke from the exhaust and sluggishness at high rpm are surefire indicators a new air filter is needed.

Other culprits that could be stealing power away from your engine:

■ Exhaust manifold leaks resulting in lack of boost on turbocharged engines is the root cause of dramatic power losses. It may also be due to mechanical problems in the turbocharger, but this is rarely the case.

■ A laboring engine that should be running free and easy indicates dragging brakes or an open-center hydraulic system not unloading.

■ Low engine coolant temperature will cause noticeable power loss. The cause? Usually a thermostat stuck open. Also, some engines use air shutters on the radiator to regulate cooling airflow. If not fully closed, the engine fails to reach normal operating temperature in cold weather.

■ Dirty or damaged injectors and high-pressure pumps can cause rough idling or poor performance under load. For example, if one or more cylinders miss while idling and then start firing when speed and load are increased, it's a good bet the injectors in those cylinders are defective. A defective injector will atomize fuel sufficiently to fire when the volume of fuel is increased. This is accompanied by blue or white smoke from incomplete combustion.

If you suspect the injectors are not performing properly, remove them and have them checked out and rebuilt if needed. Pumps are seldom a problem unless they are damaged by dirt or water in the fuel.

■ Inferior fuel can also result in low power, hard starting, excessive exhaust smoke, and rough idle.

■ We'd rather not find internal defects when diagnosing power loss. But when the fuel injection system checks out and the engine still starts hard and runs rough or misses at all speeds, then suspect the worse. A compression test will tell for sure if rebuilding is needed. ■

Steven Parks was the lead mechanic for a construction firm where he kept 'em running year-round.

Photograph: Ron Van Zee

Test Glow Plugs for Easier Winter Starts

Check each plug once a year to discover those that are burned out

By Keith Berglind

Glow plugs are the most common cold-start device for diesel engines. And being a small, simple electrical device, they are prone to failure. Since the V-8 diesel in your pickup has eight glow plugs, how do you know when first one, then another, fails?

The first symptom of glow-plug failure is, of course, hard starting. If you're reduced to plugging in the block heater or to using starting fluid even on warm days, it's time to check the glow-plug operation.

PERFORMING AN IN-PLACE TEST

I prefer to pull the wires from each glow plug and use an ohmmeter to check the internal electrical resistance between the post and engine block. For the 6.9L diesel Ford engine in my truck, the ohmmeter reading should be just under 1 ohm. When you get a reading of 0, the plug is burned out. When the reading is high, such as 10 or 20 ohms, the plug is shorted out.

To get an exact value for your test, first measure a new glow plug. This is a lot faster than trying to find a shop manual with specifications.

Once the wires are off, I disconnect the individual feed wires from the relay on the fender and check each wire for continuity and resistance. It's not unusual to find a broken wire, possibly caused by a shorted glow plug.

Another method is to engage the relay with a jumper wire, then test each wire end for proper voltage. Be sure to remove the jumper wire before connecting the glow plugs. Plugs that stay energized quickly burn out.

I like to do such in-place plug checks once a year. Any glow plug with a bad reading gets replaced. And any glow plug that tests wrong (high or low) should remain disconnected until it's changed. The last thing I want is to have the tip come off.

Some mechanics prefer to remove the glow plugs and test them on the bench with a battery. The idea is to see if the plug lights up and glows.

This method works if you're careful and know what you're doing. We tend to think all engine components run on 12-volt power. Not so. Before testing any glow plug, clean off the base and find the voltage rating.

TOO MUCH JUICE

For example, the glow plugs on my truck were stamped 11.5 volts. If the system is older and has a resistor in the feed line, the rating may be 9.5 volts. If you connect a lower voltage plug to a fully charged battery (13.2 volts), your glow plug may light up rather dramatically. That's why I don't like the bench test.

One common tractor resistor was on the dash; it consisted of a 1-inch-diameter cover with holes. Behind that cover you could see the resistor coils that glowed when the preheat circuit was engaged. When the device was glowing red, the engine glow plug was supposed to be hot, and it was time to hit the starter button.

This design had a series circuit. So if a wire broke or the glow plug died, the dash unit didn't heat up.

CHECK TIMER, AS WELL

Diesel systems in modern pickups are timed for about 11 seconds.

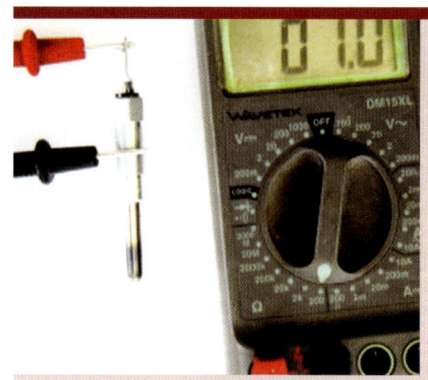

Use an ohmmeter to check the continuity and resistance value of glow plugs. As a guide, first test a new glow plug to establish a base resistance value, which for the plug above is 1 ohm.

When the red light goes out, engage the starter. When there is a failure of the timer-controller system, the warm-up time may be wrong. Too short a time period and the engine won't start. Too long a burn and plugs may burn out.

Sometimes the timer engages again after the engine starts. If this lasts for only a second or two, it hardly justifies an expensive new timer device. ∎

Photograph: Keith Berglind

Slow Cranking?

It's usually not the starter's fault; more often it's a bad ground path to the battery

By Dave Mellow

Illustrations: Paul Bridgford

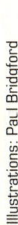

The grounding cable is connected to the battery box brace. Between that brace and the starter are as many as six interfaces or connections that may be rusted or corroded, which increases resistance.

When working around engines that won't crank, you often hear someone comment, "The starter is dragging." While there are definitely a few starters around that need an overhaul, in the vast majority of cases the more likely cause of slow or no cranking is a faulty connection.

But before you grab a wrench and tighten the battery cable's lugs thinking the job is done, be warned that grounding the pathway from battery to starter crosses many other connections. Often a battery is grounded to its box. This box is bolted to framework that supports the instrument panel or maybe the fuel tank. The metal is fastened to the main crankcase or engine cradle, which is connected to the starter through the clutch housing.

After self-propelled machinery gets a few years old, rust and corrosion creep into these various metal interfaces. The result is less-than-ideal ground paths between the battery and starter. Then, too, the starter can get loose over time, which allows grease to seep between its connection to the clutch housing. Lubricant mixed with dust makes a great insulator.

FIRST THINGS FIRST

The first tact to take when trying to cure a slow starter is to look for poor connections between the battery and starter. That pathway should be followed and connection points inspected, removing rust and grease or tightening fasteners. Even relatively new machinery can have poor pathways caused by paint, mill coatings, and lubricants.

Work to get a solid footprint (or contact area) on all mountings and connections. There shouldn't be any lock washers or empty space between wire lugs and terminals. And lugs should be positioned to get maximum area-to-area contact.

Often on older machinery there is little that can be done to overcome accumulated electrical resistance from deteriorated metal-to-metal connection. When faced with such a situation, it's best to establish a new grounding path from the battery direct to the starter, for example.

GO WITH AN EXTRA-LONG CABLE

Replace the ground strap with a cable long enough to reach the starter housing bolts. While you're at it, buy a cable that has that little auxiliary 10-gauge wire attached to its terminal. Run that extra wire to the instrumental panel to ground it for good measure. ∎

Removing the starter from its engine housing reveals its working end, the bendix, and a likely location of trouble for a unit that is not fully engaging the engine's flywheel. The coiled spring at the right forces the bendix gear away from the flywheel when you let up on the ignition switch.

Sluggish Starter

Expert explains how to examine and recondition a slow-spinning starter

By Ron Van Zee

Jim Friedrich bemoans the lack of respect starters receive. They are often overlooked at maintenance time and are the last component examined when hard starting occurs. The battery, cables, and connections are more often examined when a tractor isn't spinning over at a brisk pace, Friedrich says. Not that these components are without blame.

THE ELECTRICAL SYSTEM

Friedrich advises inspecting the entire electrical system servicing a starter. Give the battery a good charge and then check its strength.

Clean battery posts and the inside surfaces of cable clamps to make sure solid connections are made. And while you're about this job, remove, inspect, clean, and retighten all connections from the battery to the ignition switch, solenoid, and starter. The point here is to eliminate these components as the source of starting problems.

Friedrich encourages replacing old, frayed, and worn cables running from the battery to its grounding point and to the starter. The cost of doing so is cheap and the results can be remarkable.

THE STARTER

If, after the electrical system is inspected and repaired, a tractor is still turning over slowly, the finger of blame likely points at the starter, particularly on older tractors that have a lot of hours on them.

Friedrich recommends trying to start the tractor again. The purpose here is to listen only to the starter's operation. "We're not concerned if the engine starts, as that depends on the health of the fuel system (fuel or injector pump), cylinder compression, or valve timing," Friedrich says.

Instead, monitor the speed at which the starter turns the engine over. If it is slow, this might indicate the starter's brushes are in disorder or the commutator is scored.

At this point in your troubleshooting, remove the starter from its engine

Jim Friedrich operates a garage that specializes in tractor repair in Lake Elmo, Minnesota. Increasing interest in tractor restoration inspired his business's name, Rust In Peace Tractor Resurrections. Contact Friedrich at 651/770-7699.

mounting and either take it to a professional or repair it yourself. The following repair advice from Friedrich mostly pertains to starters on older tractors.

Your problem may be found inside the starter. You can possibly inspect brushes and the commutator through access holes found at the rear of the case prior to disassembling the starter. But removing the armature allows you to completely clean the unit's exterior, removing all dirt and grease from the case as well as the bendix. The bendix comprises

Photographs: Ron Van Zee

the spiral shaft, gear, and coiled spring that reside inside the engine.

Before disassembling the starter, clean the component's exterior to avoid transferring grime to interior components. Plus, a clean starter is easier to disassemble, Friedrich says.

THE BENDIX

After cleaning, examine the bendix to see if it operates smoothly. Look at the engaging gear to see if its teeth are worn or broken off. Damage will likely warrant replacing the bendix. If a replacement part isn't available, you may have to resort to buying a new or reconditioned starter.

THE ARMATURE

Next, grab the armature shaft and try wiggling it side to side. This may be an indication of front bearing wear. A new bearing is likely needed to prevent the armature from dragging on the field windings.

Next, remove the armature from the starter's case for inspection and cleaning. Inspect the brushes and their pressure springs for wear and breakage. Worn brushes and broken spring will need to be replaced.

THE COMMUTATOR

Examine the commutator to see if it is dirty, burned, or rough. Clean the commutator well, Friedrich says.

To do this lightly clamp the armature in a vise. Using a strip of fine sandpaper or emery cloth, lightly polish the commutator. If the part appears to be scored or burned, you may have to have it turned on a metal lathe to get it back into operating condition.

Assemble the armature in its case, reinsert the starter in the engine, and see if your labors made a difference starting the engine.

If your work didn't improve the starter's performance, you will have to take it to a professional mechanic for a rebuild or replacement. ∎

The point at which all a battery's power is put to work turning an engine over is at the bendix. Notice the worn and chipped teeth on the bendix in this photo. Complete teeth on the gear are crucial to transmitting all the power from the starter's armature to the engine's flywheel.

A starter's brushes involve the four copper conductors (located around the hole in the center) that surround the commutator. The coiled springs to the side of the brackets that house each brush push their respective brushes down to make contact with the commutator.

Brushes contact the starter armature's commutator (the copper cylinder right of the shaft). This contact is crucial because it transfers needed electricity. The transfer causes the armature to rotate, which engages the bendix gear and turns an engine over.

Stemming Oil Consumption

A step-by-step guide to replacing valve stem seals

By Steven Parks

You've probably seen the effects of defective valve stem seals and not known it. Engines blowing a cloud of blue smoke may be suffering from defective seals, especially if the engine has normal compression and smokes when restarted after sitting 20 to 30 minutes.

Valve stem seals control the amount of oil allowed between the valve stem and guide. In doing that job they have an impact on oil consumption – so much so that defective seals can boost oil use by up to 70%.

These seals are typically made of four different types of synthetic materials. The quality is based on their ability to handle temperature.

Nitrile: Has the characteristic of becoming hard and brittle when exposed to temperatures over 250°F. Many OEM stem seals are made from this cheaper material and may fail prematurely.

Polyacrylate: A step above nitrile both in temperature tolerance, (they're able to endure 350°F. before becoming hard and brittle), as well as in price. Polyacrylate seals cost twice as much as nitrile.

Silicone: Withstands temperatures from 330°F. to 375°F., depending on the grade of material. The cost is four to five times greater than nitrile.

Viton: It tolerates temperatures up to 450°F. with good flexibility and wear resistance. Although Viton costs about 10 times more than nitrile, it's still the material of choice for most heavy-duty uses.

INSTALLATION TRICKS

It is possible to replace the valve stem seals on many engines without removing the cylinder head. Pressurizing the cylinder with compressed air through a screw-in fitting that replaces the spark plug will hold the valves in place. Then you can compress and remove the valve springs and change the seals. Overhead valve engines that use pushrods to activate the valves are good candidates for this procedure.

This is not the case with overhead cam engines, however, because of the greater complexity of the cylinder head. And it's tricky to hold the cam chain in place while the force of the compressed air acting on the pistons rotates the engine crankshaft. You may prefer to remove the entire cylinder head on overhead cam engines.

When using the air pressure method on overhead valve power plants, include these basic steps to valve stem seal replacement:

- Thoroughly clean the engine as well as all the area under the hood.
- Remove a battery cable for safety.
- Remove all spark plugs, marking the wires for correct reinstallation.
- Remove the valve cover, rocker arms, and pushrods.
- Inspect the ends of the valve stems for mushrooming and sharp edges that may damage the new seals during installation. If these problems are found, remove the entire cylinder head so the valves can be properly reconditioned.
- Be sure that all head bolts are in place and properly torqued. Then install your air fitting adapter into a spark plug hole and pressurize it.
- Compress the valve spring on the pressurized cylinder with an appropriate tool. Next, remove the retainer and spring. Set them aside on a clean lint-free cloth.
- Remove the old valve stem seal, wipe the area clean. Install the new stem seal according to the manufacturer's directions.
- Install the springs and retainers. A bit of stiff grease will hold the retainers in their groove while you decompress the valve spring. ■

Valve stem seals are available for a wide variety of valve types.

Use the appropriate tool to remove the valve spring and retainer.

Install the new seal according to its manufacturer's instructions.

Photographs: Steven Parks

Why Batteries Go Bad

Sulfate attacks idle batteries not kept charged

By Dave Mowitz, Machinery Editor

Idle batteries can become permanently dead batteries, creating a disposal problem. To get the full life out of batteries, keep them fully charged during the winter months.

Winter is the worst time of year for batteries. Not because they are put to hard use turning cold engines over.

"Often it's not the use but the disuse that kills batteries before their time," says Andy Anderson of BatteryStuff.com.

Batteries sitting idle for long periods of time sulfate. Anderson explains that the sulfuric acid in batteries not kept fully charged tends to stick and stay on lead plates. After a short time, sometimes less than a month, these plates become so coated with sulfur they can't be recharged.

The key is to keep batteries fully charged, as units kept at 75% or less than their capacity are vulnerable to sulfate. To do this, use one of the new generation of smart chargers that sense charge levels and replace energy only as needed.

Other winter storage tips for batteries include:

■ Take batteries out of vehicles idled for long periods of time.

■ Store these batteries in a dry and warm (not hot or cold) location. Particularly avoid hot locations because batteries discharge as their temperature increases.

■ Keep batteries on a wooden shelf. Batteries stored on concrete can lose their charge in a few weeks. ■

CLEAN FOR LONG LIFE

Add to your winter to-do list a maintenance chore that pays huge dividends when it comes to prolonging battery life: cleaning and coating terminals. Outside of the time consumed, this is a cheap maintenance task since it requires only baking soda, water, some caulk, and grease to complete.

Remove battery cable connections. Mix a couple of tablespoons of baking soda in water. Dip a fine wire brush in this solution and apply liberally to the battery terminals and couplers. Let the baking soda work a bit to neutralize any corrosion, while brushing away contaminants. Finish the job by washing all parts with clean water.

To prevent future corrosion, apply a thin bead of silicone caulk at the base of battery posts. Place a felt battery washer over the caulk. Now coat the felt washer with high-temperature grease or petroleum jelly. Also, coat the exposed cables ends with grease. This covering prevents acid gases that escape from the battery from condensing on metal parts, which is the source of most corrosion. ■

RATING BATTERY POWER

Here's the lowdown on those sometimes confusing battery ratings:

■ Cold cranking amps (CCA) represents the amps a battery delivers at 0°F. for 30 seconds while staying at a 7.2-volt output or higher. A high CCA rating is an indicator of a battery's capacity to work in cold weather.

■ Cranking amps (CA) is determined while the battery is being drawn down at 32°F. This rating is also called marine cranking amps (MCA).

■ Reserve Capacity (RC) lists the number of minutes a fully charged battery discharges 25 amps at 80°F. until it drops below 10.5 volts. This makes RC a particularly crucial rating. A term related to RC that's no longer used is hot cranking amps.

■ Amp hour (AH) is a rating typically listed on deep-cycle batteries such as those used in tractors and combines. If a battery is rated at 100 amp hours, it should deliver 5 amps for 20 hours, 20 amps for five hours, and so on. ■

Photograph: Dave Mowitz

Putting Putty in its Place

The key to body filler longevity, regardless of the filler being used, is applying the putty to bare metal to strengthen adhesion.

Quality fillers do different jobs

By Dave Mowitz

I'm not a great fan of using auto body putty as a fix-all for any sheet metal restoration challenge that comes down the pike, particularly for filling in deep dents, nicks, or gouges. That's because any vibration can cause the body putty to crack or pop off metal. And tractors vibrate a lot more than cars. Which makes them poor candidates for the wholesale use of putty – a common practice in auto repair.

"You're always better off pounding out deep dents or replacing heavily pitted parts with new metal," says veteran restorer Jeff Gravert of Central City, Nebraska.

A $50 to $90 investment will buy you a good set of auto body tools. For example, Eastwood Company sells a seven-piece set of professional-grade hammers and dollies for $79.99 (call them at 800/343-9353 or visit www.eastwoodco.com).

GREAT FOR SMALL JOBS

Body putty, however, when applied in thin layers, can be a godsend when filling small imperfections and pits or hiding a patch job. Naturally, the key to any successful putty job is preparation.

Putty, like paint, adheres better and stays intact when applied to bare metal. So make sure you completely remove all oil, grime, solvent, rust, and paint from the surface.

Now, in the good old days, your choice of body putty was limited to plastic fillers. Not anymore! I checked in on Evercoat (513/489-7600 or www.evercoat.com), a source for professional auto body supplies, and counted no less than 15 different filler products.

Offerings range from a lightweight resin filler called Rage Xtreme (guaranteed to be pinhole free) to Metal-2-Metal. The latter incorporates fine aluminum particles, and Evercoat brags it's the "nearest thing to lead."

Another Evercoat innovation, Fiber Tech, is formulated with Kevlar, which gives it superior strength.

WHAT'S THE BEST FILLER?

When in doubt as to which putty to use, the best bet is to start with a basic Bondo-like product, taking care to apply it properly and in layers. Remember to sand between layers and remove all dust before applying the next coat.

But don't be completely put off by alternatives. Evercoat's Metal-2-Metal virtually eliminates corrosion under patch areas, making it ideal for use on surfaces exposed to water (from radiators) or fluids (like fuel). I worked with a sample of metal-filled putty and was impressed with how well it sanded out.

Another filler product worth taking a look at is a liquid-thin glazing putty that fills in all pinholes and microscopic seams.

BUY A SPREADER SET

I do recommend buying a spreader set consisting of different types of flexible spatula-like trowels. Some sets will come with a plastic mixing board. But I have found that using cardboard covered with aluminum foil works just as well, and it costs less. And when you're done, just toss the foil. ■

Photograph: George Ceolla

Spark Plug Thermometer

The orange arrows illustrate how the ceramic insulation around the firing pin affects how far heat must travel to reach the cylinder head, and thus be dissipated.

Hot spark plugs literally operate hotter

By Dave Mowitz, Machinery Director

A common misconception about the terms *hot* and *cold* as they apply to spark plugs is that hot refers to higher levels and cold refers to lower levels of voltage.

Turns out the terms are all about heat – or the lack of it.

A hot plug operates hotter than a cold plug, explains Sam Steadman of Autolite, thanks to the fact that the length of its center electrode or firing tip is covered with more ceramic insulation.

"Heat range is a measure of how fast a plug's firing tip dissipates combustion heat," he says.

A cold spark plug has less of the length of its center electrode covered with insulation. This allows combustion heat to easily travel from the firing tip to the plug's shell and then into the cylinder head where it dissipates. The insulation on the tip of a hot plug is longer, which requires combustion heat to travel farther before dissipating.

Other factors affecting heat range are the thermal characteristics of the insulator and the bond of the insulator to the plug's shell.

A BALANCING ACT

The important balance between hot and cold is crucial to tip performance, Steadman says. The firing tip has to operate hot enough to burn off combustion deposits. But a tip must stay cool enough to keep from preigniting the fuel-air mixture as well as to prevent eroding the side electrode.

Plugs are designed in many ranges to suit a variety of applications. Heat ranges can change a plug's operating temperature from 70°F. to 120°F.

"If you are unsure of the correct heat range to use, always start with a colder spark plug," Steadman recommends. "If it is too cold, the plug will eventually foul and misfire. While this results in rough operation, it won't hurt the engine."

Selecting too hot a plug can lead to preignition and detonation that can severely damage an engine. ■

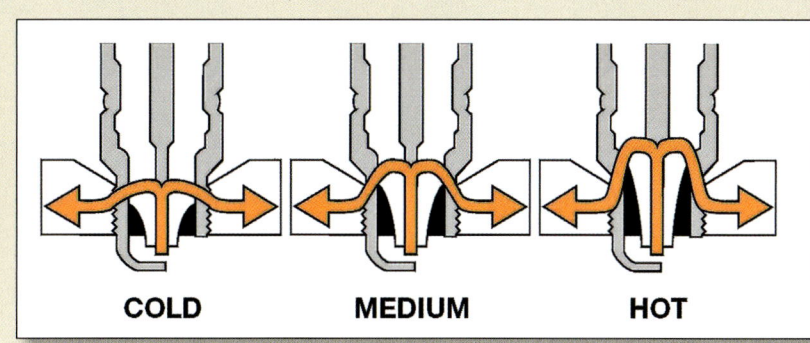

COLD   MEDIUM   HOT

GAUGE ENGINE HEALTH BY PLUG COLOR

The color and condition of a plug can reveal much about what is going on inside your engine. For example, a plug coming from a properly operating engine is grayish-tan to white. Don't be alarmed if the plug is pinkish-red because this comes from using additives in unleaded fuel. Appearances that warn of engine problems include:

■ An insulator tinted charcoal with a firing tip that is damp with gas indicates a faulty choke, overly rich fuel mixture, ignition problems, leaking fuel injectors, or too low a plug heat range.

■ A firing tip covered with black soot can indicate too cold a plug heat range, an improperly adjusted carburetor, or malfunctioning choke.

■ When the ceramic tip, center, and electrodes are coated with a black, oily substance, then oil is entering the combustion chamber, indicating worn rings, valve guides, or valve seals.

■ A cracked, chipped, or broken insulator is caused by low-octane fuel or overadvanced timing.

■ Electrodes that have rounded edges, are eroded, or excessively worn away should be replaced. ■

Any air conditioning system that is being converted to use R-134a refrigerant should receive a detailed inspection. Plus all of the old oil used with R-12 refrigerant must be removed.

Alternative Refrigerants

The solution to replacing banned air conditioning refrigerants
By Steven Parks

If you haven't already encountered problems recharging the refrigerant in your tractor's or combine's air conditioning system, then be prepared for this chore to be anything but swift and simple in the future. R-12, the standard refrigerant for decades, hasn't been produced since 1995 when environmental laws ended its legal production. Existing stocks of the stuff are nearly depleted, so very soon R-12 won't be available at any price. Any company that tells you differently is likely stretching the truth and trying to sell you an alternative to R-12. These replacements are blends of several different refrigerants. A few that have failed to gain EPA approval are downright dangerous, as they use flammable hydrocarbons like propane or butane in the refrigerant mix.

REPLACEMENTS TO ASK FOR

When shopping for a replacement for R-12 refrigerant, ask some questions about the product. Does it contain flammable components? Is it nontoxic and safe to use?

Does the product contain R-22, which requires barrier-type hoses? Is it accepted by the EPA under SNAP (Significant New Alternatives Program; www.epa.gov)? Is this replacement approved by the manufacturer of your air conditioner? Find out if it will be available in the future, and if it offers any advantage over R-134a.

R-134a is the only R-12 replacement accepted by the mobile air conditioning industry. The substance has been used in new vehicle systems since 1992 and is widely available to certified air conditioning technicians at less than $5 per pound. Extensive testing shows that it's safe and effective.

But you can't just grab a can of R-134a and squirt it into your air conditioning system. For starters, using R-134a requires some prep work. This begins by following the guidelines for conversion outlined by your manufacturer. That information takes precedence over these general guidelines.

Also, avoid making the conversion to R-134a unless you know what you're doing. An untrained person may inflict self-injury or ruin expensive parts. Do the training, get the certification and tools, or hire someone who is qualified.

If there is any R-12 or other refrigerant left in the system, it must be recovered with special equipment. It's illegal and dangerous to vent refrigerant directly into the atmosphere. Recovery equipment starts at $600. So unless you have a large fleet of vehicles to convert, you'll want to hire someone to do this work.

If you are doing the conversion, start by making a careful inspection of the vehicle's system. Is the system in good shape or is it plagued with worn parts, leaks, and electrical malfunctions? If so, squirting it full of new R-134a won't fix anything. Any repairs must come prior to a retrofit.

MAY NEED A NEW CONDENSER

Switching over to R-134a may also require a new condenser. This is because R-134a runs at slightly higher pressures and temperatures than R-12, and requires a larger condenser to vent the same amount of heat. This means if your existing system has marginal cooling performance using R-12, you may be disappointed after converting to R-134a.

Before installing new refrigerant, it will be necessary to drain the old oil from your system's compressor. This is necessary because the mineral oil used with R-12 will not work with R-134a. ∎

Photograph: Ron Van Zee

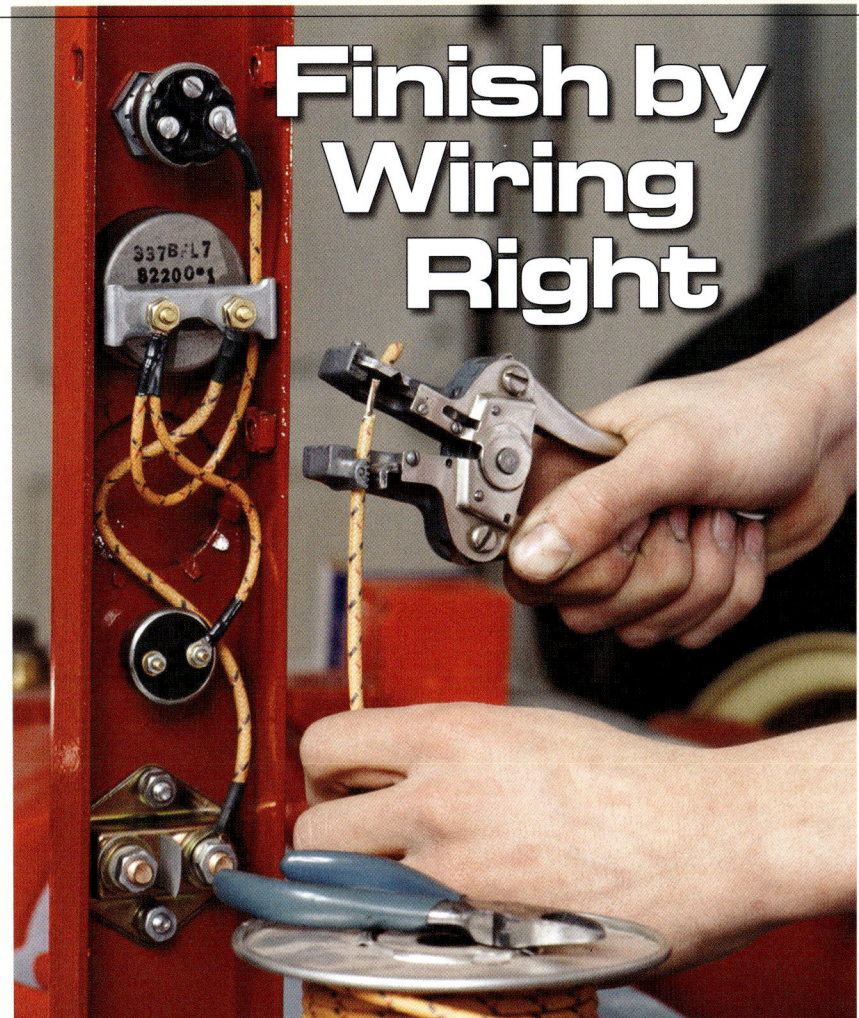

Finish by Wiring Right

Photograph: Ron Van Zee

When sizing wiring to a particular job, bear in mind that its carrying capacity must match the total load for a particular circuit. The greater the load it has to carry and the longer the path it must run, the greater the resistance in the wire will be. This resistance can be reduced with a larger diameter wire.

Dirty and plugged filters are the leading culprits in stealing power

By Steven Parks

As tempting as it may be, don't scrimp on rewiring. Compared to engine parts or a paint job, rewiring expense is nominal. Even if old cloth- or plastic-covered wiring looks sound, it may be long past its prime for transferring electricity efficiently. Wiring terminals and connections that are corroded, rusted, or have disappeared entirely not only compromise engine operation, but also pose a fire hazard.

SELECTING THE PROPER WIRE

The best advice is to go with entirely new wiring. The good news is there are a number of firms that offer a wide selection of cloth wiring to match the original equipment on your vintage tractor.

Selecting a wire size depends on a wide variety of conditions such as the voltage and amperage expected, where it is used (transferring power from the battery to the starter vs. from a magneto to a spark plug), and the length of its run. Size difference reflects the fact that as electricity moves through a wire it builds up resistance. This resistance, in turn, creates heat.

Now, the longer the wire and the greater the load it is carrying, then the greater the resistance. Too small a diameter wire in certain situations can generate enough heat to make a wire's covering melt.

When it doubt, refer to your service manual for the proper wire size recommendations. If none are provided, then refer to the wiring chart in an automotive guide. Many public libraries carry Chilton Guides, a great source of information.

And then there is my favorite source of restoration information – other mechanics. Get online and go to a Web site that caters to the brand of equipment you're working on. I'll guarantee you a whiz-bang mechanic at that site will respond to your discussion group plea for advice.

INSTALLING WIRE

After buying the proper size wire, take a bit more care to install it correctly. Use new terminal ends that are properly crimped or soldered to assure a solid connection. Thoroughly clean original terminal ends to reduce resistance. And tighten locking washers, when used, so they make solid contact. I like to use terminal boots or a covering on all connections that are exposed to weather.

Other installation rules:
- Avoid make sharp turns in wiring runs.
- Keep wiring away from high heat surfaces, such as a manifold, or from areas where it might be exposed to oil or fuel.
- Always use a rubber grommet to protect wires passing through sheet metal.
- Run multiple wires in a harness or loom.
- Secure wiring using wire clamps or clips. ∎

Massive rigs like this air seeder can demand as much as 60 gallons-per-minute hydraulic flows to accurately propel seed from the cart to the drill. And that accuracy can be quickly jeopardized if a system is not properly adjusted to provide the extra flow and pressure to other components in the system such as marker arms and lift cylinders.

Unraveling Hydraulic Headaches

The key is to understand and pamper your hydraulic systems

By Dave Mowitz, Machinery Editor

I had my first lesson in hydraulics at the ripe age of 15. My dad was shopping for new horsepower, and one of the local dealers brought out a tractor to show off.

During the demonstration, Dad wanted to see if the tractor could lift our eight-row mounted planter. It couldn't, and the dealer didn't make the sale.

Later that day, the dealer called to explain that the flow control for the three-point had been closed down by half. The explanation still didn't convince my dad to buy, but it gave me a new insight into hydraulics. "You mean you can actually control the speed of a three-point?" I wondered.

Sounds simplistic, I know. But the hydraulic experts I interviewed for this article point out that misadjusted flows are still the number one mistake farmers make with hydraulic systems.

That oversight has been amplified by skyrocketing demands that implements make on hydraulic systems. Planters, in particular, drink hydraulic fluid like a thirsty cow. Plus they demand precise pressures to regulate seeding or fertilizer rates.

Manufacturers have responded to those demands. A decade ago, 100-plus-hp. tractors boasted hydraulic flows of 20 to 25 gallons per minute (gpm). Today, that figure has jumped to 40+ gpm, and four-wheel-drive rigs can put out 60+ gpm.

Besides being more powerful, today's hydraulic systems are smarter, as well, with "advanced electronic controls that make hydraulic allocations easier. They are also more efficient," says Gary Dostal of John Deere. "And most hydraulic functions can be controlled in the cab."

Yet, it is still crucial to understand the hydraulic system on your tractor – old or new – to meet increasing demands. That understanding starts with identifying the difference between open-center and closed-center hydraulic systems (see below).

OPEN-CENTER HYDRAULICS

Up until the 1970s, nearly all tractors employed open-center hydraulics. This system is still common on many new tractors up to 80 hp. The approach derives its name from the fact that the system's control valves are always open to the reservoir when in the neutral position. As a result, the pump is constantly sending oil through the valves and back to the reservoir.

Such systems are reasonably priced, tried-and-true, and more than adequate when it comes to intermittent use. Open-center pumps, which are generally gear designs, are speed sensitive: the faster you

Photographs: Ron Van Zee

OPEN-CENTER SYSTEM IN NEUTRAL

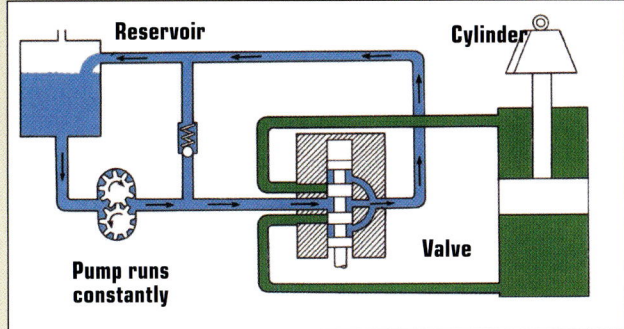

In an open-center hydraulic system, the hydraulic pump runs constantly when a valve is in the neutral, or open position. This circulates oil through the valve and back to the reservoir. When the valve is actuated, oil flows to the load, illustrated as a cylinder pushing a weight above. Oil on the back side of the cylinder's piston returns through the valve and then to the reservoir, feeding the pump.

CLOSED-CENTER SYSTEM IN NEUTRAL

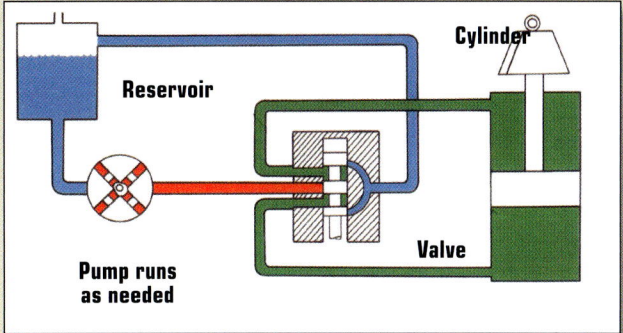

Closed-center hydraulics employ a variable displacement pump that idles when a valve is closed, or in the neutral position. Standby pressure is maintained by a pressure-regulating valve. When the valve is opened, the drop in the pressure line prompts the pump to send oil flowing. Compensation of pressure and flow ensures that the exact pressure and flow needed to move the load are provided.

run the engine, the higher the flow rate. This explains why you notice performance improvement at higher engine speeds.

In the 1980s, hydraulics engineers started using closed-center hydraulics, featuring variable-displacement piston pumps on tractors. The valves on closed-center systems remain closed when in neutral. And their pumps are at idle until called on for flow. At this time, the pumps deliver only enough flow to maintain standby pressures of approximately 2,900 psi. When a valve is activated, internal sensors detect a drop in system pressure and promptly crank up the pump.

Recent improvements to closed-center systems include the use of pressure and flow compensation (PFC). "This feature provides oil flow to multiple functions on the tractor simultaneously," says Paul O'Hara of AGCO. "It helps equalize the demand on a system, delivering more precise flow and pressure."

CONTROL THE FLOW

Whether it's open- or closed-center hydraulics, you must still pay

5 WAYS TO BOOST HYDRAULIC LIFE

Want to dodge major breakdowns while adding life to any hydraulic system, new or old? Then practice these five preventive maintenance chores offered by hydraulic experts.

1 Change hydraulic oil once a year – at a minimum – to remove contaminants. Water, in particular, is hard on wet brakes and powershift clutch packs on newer tractors in storage. The best time to change hydraulic oil is at the end of the season and before contaminants suspended in oil settle out. "The first oil change on new tractors is critical as this removes debris deposited by engine break-in and factory machining," says Tom Kane of Kubota.

2 Take the temperature of a system. Warm pumps, motors, hoses, or cylinders are normal. But if any component is excessively hot, there's a problem. Operating with high oil temperatures is like running an engine with high coolant

temps. Damage is guaranteed. The most likely causes of high temperatures? A poorly adjusted or overloaded system, or a dirty or plugged oil cooler.

3 Check hydraulic oil as often as you do engine oil. Do both once a day – at least – or after every 10 hours of use.

4 Always drain and change hydraulic fluid on just-purchased used machinery. Expect that any oil in such equipment is contaminated and of inferior quality. If, during draining, you find the oil is badly contaminated, then flush the entire system on tractors and combines.

5 Keep the system "clean, clean, and clean," emphasizes Gary Dostal of John Deere. "Use a clean cloth to clean tips and receptacles before coupling . . . and don't use that rag again. Be sure to keep oil and components clean. I will guarantee this will increase a system's life." ■

Regardless of improvements in covers and caps, the number one source of dirt contamination of hydraulic systems is still selective control valves. And the number one cause of that contamination is not cleaning couplers.

at the marker and motors. "The requirements of these applications are different," Renaud says. "If all the control valves are wide open, oil will flow first to the lower pressure of the marker arms (some 3 to 5 gpm at 500 psi), then to the motor (4 to 10 gpm at 1,000 psi), and finally to the lift cylinders (8 to 13 gpm at 1,800 to 2,500 psi)."

To avoid cross talk, calculate how much flow each component requires. Then adjust the respective valve to that application. Do this by closing down all valves. Next, starting with the largest application (the lift cylinders in the previous example), open the control in increments. Operate the cylinder after each adjustment.

"When the lift rate is satisfactory, you hit the right flow. Now go on to the next highest requirement (the orbit motors in the previous example)," Renaud says.

ORBIT MOTORS AT WORK

When working with a closed-center system with PFC feeding orbit motors, be sure to control the flow at the tractor rather than at the implement. If you are using open-center hydraulics to power orbit motors and it's overheating the system, consider running smaller orbit mo-

attention to regulating flows. "Too often farmers set flow control valves wide open," explains Dan Renaud of Case IH. "This can cause cross talk between valves when more than one valve is activated."

Oil will always take the path of least resistance. If all the valves on a tractor are set wide open, the application requiring the least pressure will always be serviced by the system first.

This explains why, when a planter and its marker arms are lifted at the same time that the orbit motors are operating, there is a pressure spike

CARE AND FEEDING OF COUPLERS AND HOSE

What is the number one item to break down in a hydraulic system? The unanimous choice of the hydraulic experts in this story is hoses and couplers. Here are their recommendations for improving the care of these components.

■ Check all hydraulic lines and connections once a week during the height of fieldwork.

■ Eliminate sharp, 90° turns and loops in lines since they can restrict flow and promote breaks.

■ Avoid twisting hoses. Twisting a hose's wire reinforcement just 5° reduces hose service life by 70%.

■ Eliminate pinched, dented, or obstructed lines. They cause foaming, overheating, loss of power.

■ Install hose with some slack in the line. Taut hoses will strain in operation, which causes them to bulge and eventually weaken under pressure.

■ Avoid mixing low- and high-volume couplers. The industry switched to high-volume couplers several years ago to take advantage of higher pump output. Low- and high-volume connections look the same and will work together. But mixing couplers can reduce flow or stop oil from moving altogether.

■ Match hose size (listed as inside diameter or ID) to flow volumes. A hose with too small an ID causes oil turbulence, pressure drops, overheating, and tube damage. ■

tors that run at higher pressures to reduce heat.

"Smaller motors won't steal flow away from other functions of the tractor," explains Dostal.

Overheating is common when high flow demands are placed on open-center hydraulics. "Such systems typically require the use of special control valves or flow diverters to direct, orifice, and regulate flow for proper operation," Renaud says. "Even with best-case scenarios, excess heat buildup while running orbit motors remains a major obstacle, requiring additional oil coolers to protect the tractor."

When working with power-beyond components on large implements, be sure to feed such systems from the full-flow port found near a tractor's valve banks. Selective control valves won't provide enough flow or pressure for such components to operate properly, warns Todd DeBocker of New Holland. "Power-beyond needs full flow of oil from the tractor since it acts as an extension of a hydraulic system's valves."

Your dealer will have to install a full-flow port if your tractor isn't equipped with that feature.

SYSTEMS HAVE TO BE THE SAME

Be sure that valves on implements are compatible with your tractor's system. Implements with open-center valves should only be fed from an open-center system on the tractor. The exception to the rule involves closed-center systems with PFC, which can accommodate any valve type on implements.

Finally, remember that your tractor depends on hydraulics to operate clutches, brakes, steering, and other systems. So don't count on that flow to feed extra cylinders or motors. The exception to this rule is on tractors that have a separate pump that serves only on-board functions. ∎

NOT ALL HYDRAULIC OILS ARE MADE ALIKE

Sticker shock from buying high-quality hydraulic transmission oil is understandable. After all, the good stuff can be double the price of generic brands.

But fight the temptation to buy cheaper fluid or skip an oil change. It will cost you poor hydraulic performance today and thousands of dollars in future repairs. "A major breakdown of a hydraulic system often involves the pump," says Tom Ogle, McCormick Tractor. "And when the pump goes, you are dead in the water because today's tractors and implements are completely dependent on hydraulics to operate."

High-quality fluids are created from high-quality base oil stocks and enhanced to do several jobs at once under high pressures and temperatures. "When oil additions are depleted and lose their protective properties," says Tom Kane of Kubota Tractor, "you can start to see excessive wear and corrosion."

MUST-HAVE ADDITIVES

Some enhancements are:
- **Viscosity enhancers.** Viscosity ratings measure oil's thickness at a given temperature. At high temperatures, fluid must be thick enough to lubricate as it minimizes leaking (from seals and joints), and at the coldest, it must be thin enough to flow readily, says Michael Zink of Quaker Chemical.

High-quality oils also extend their viscosity over a wider range of working temperatures, so they have a high Viscosity Index (VI) rating. Typical values range from 0 to 300. The higher the VI, the smaller the variation in viscosity as operating temperature changes.
- **Antiwear.** Oil has to stick to hot parts under high pressure to keep friction to a minimum. Manufacturers' additives, such as zinc dithiophosphate, form a protective barrier on metal surfaces.
- **Friction modifiers.** A must for wet-type brakes, which require friction between brake pads and disks but still need metal-to-metal protection.
- **Antioxidants.** Oil breaks down, or oxidizes, as it's used. Oxidation forms sludge, varnish, and acids. Sludge and varnish impede the function of hydraulic components, leading to problems like sticking valves. And acids eat away metals and seal materials. Oxidation is increased, sometimes very rapidly, by water, air, and other contaminants.

Antioxidant chemicals minimize the effects of oxidation and extend the life of fluid.
- **Rust and corrosion additives.** These work in combination with antioxidants to specifically retard corrosion-causing acids or air-enhanced rust.
- **Foam inhibitors.** High-quality oils are able to dissolve small amounts of air to minimize its impact. But bubbles will result if the amount of air is too great. They form when air is released from oil going from high- to low-operating pressures. Bubbles and released air create foam, which can seriously affect oil performance. Foam inhibitors eliminate foam and remove dissolved air.
- **Demulsifiers.** These improve oil's ability to resist air and water retention, too. Without them, air and water blend with oil to form emulsion (that slimy, pasty stuff), which increases oxidation and reduces oil's lubricating ability. ∎

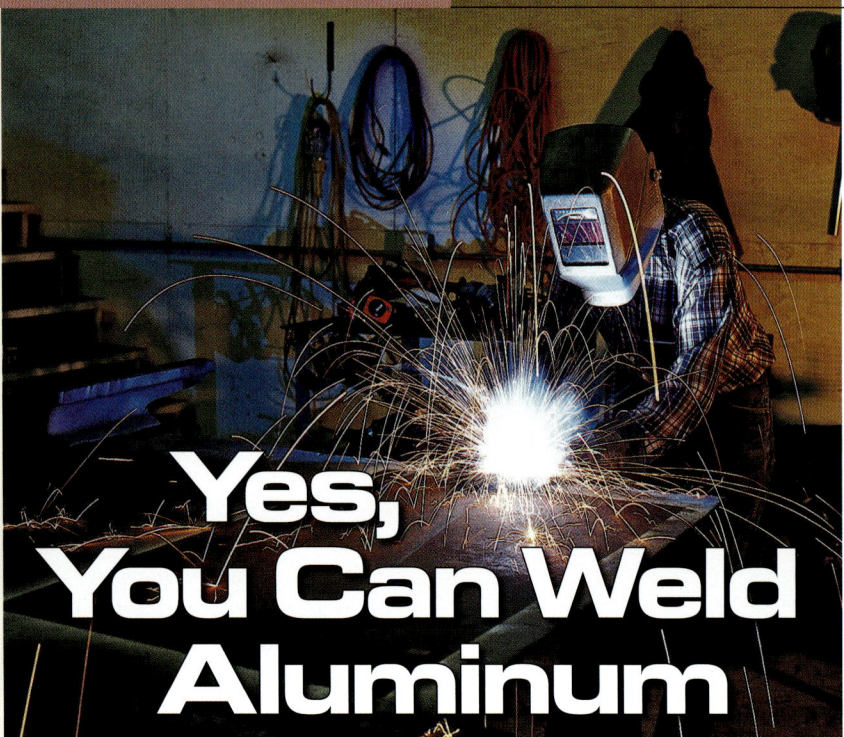

A compact 115-volt MIG welder is often all you need to tackle simple aluminum welding jobs.

Yes, You Can Weld Aluminum

The right welder and supplies are all you need

By Jim Harris

With practice, the right equipment, and proper setup, a compact MIG welder is often all you need to tackle occasional aluminum welding jobs.

Keep in mind that a 115-volt wire feeder welder can handle aluminum welding jobs that range from 22 to 12 gauge. With moderate preheating, you can probably weld as thick as ⅛ inch. Be aware that preheating should be limited to 250°F. maximum.

Another option is a 230-volt machine that can weld from 22 gauge all the way to ³⁄₁₆-inch stock. If you need to weld a broader range of thicknesses, consider investing in the 230-volt machine.

Another common decision that comes up when selecting a welder is whether you want a continuous or tapped voltage control model.

A continuous voltage control model allows you to set an infinite range of voltages within the rating of the machine. This provides more adjustability, fine-tuning, and precise control.

These features let you more easily adapt the voltage to your application and particular skill level.

TAPPED CONTROL IS CHEAPER

If you're on a budget, opt for the tapped control unit. This machine has a rotary switch with four or five fixed voltage choices. It will not give you the control of a continuous model, but it can be slightly easier to get up to speed and costs less to purchase.

For these types of welders, it is best to make welds in the horizontal and flat positions only. In general, fillet welds in lap joints are made more easily than groove welds in butt joints. Fillet welds in tee joints are preferred over corner joints.

MIG welding aluminum is different than welding steel when it comes to shielding gas requirements. For aluminum, 100% argon is the gas of choice, whereas steel welding calls for a mixed gas or 100% carbon dioxide gas. The good news is that no special equipment is needed for argon (with the exception of carbon dioxide regulators), and gas hoses can be used for both pure blends and mixed gases.

WIRE HAS TO BE DIFFERENT

You will, however, need different welding wire with aluminum. Compact MIG welders should be limited to .035-inch diameter 4043 aluminum alloy filler metal. A 5356 aluminum alloy electrode may commonly be recommended by retailers since it is a stiffer wire and can be easier to feed. However, with these types of wire feed welders, there is often not enough amperage to achieve a good weld with 5356. Even though 4043 is a softer wire, following the proper-use procedures will ensure good feedability.

You should avoid using 0.030-inch wire as it is difficult to feed through the welder. Also avoid ³⁄₆₄-inch wire as these compact welders do not typically produce enough current to reliably melt this diameter of wire. ∎

LEARN MORE

For a complete guide on welding aluminum, visit www.lincolnelectric.com.

Photograph: Doug Hetherington

Handling Hard-Facing Chores

Successfully hard-face with a bit of know-how and good prep work

By Dave Mowitz, Machinery Editor

W hen it came to welding, I was our farm's willing expert back in my youth. So when my father gathered up some cultivator shovels to be hard-faced by the local blacksmith, I grew resentful. The job didn't look so difficult. Why, all you had to do was slather some molten metal on the part's surface.

Holding back a shovel, I tried my hand at hard-facing it. The results were anything but smooth. Plus my hard-facing popped off the shovel within just a few acres of use. Real welders call that spalling.

WHAT'S IN A METAL

I never tried to hard-face again. Pity, as it is a welding job anyone can tackle if they know how to start the process of laying on hardened metals.

That begins, I found out from Tom Black, Lincoln Electric's hard-facing guru, by first identifying the part to be hard-faced. "A good rule of thumb is to assume that nothing is mild steel," Black explains. "Almost all implements are high-strength steels (either high- or low-alloy) and many are higher carbon steels."

How do you tell the difference? Use a magnet. If it sticks to the part, then that item is likely iron-based. However, a magnet won't stick to manganese or stainless parts.

Or, you can grind the part and watch what size, color, and type of spark it throws off. I could write a dissertation on sparks. Instead, I highly advise getting ahold of the books listed at the end of this story, and referring to the nifty charts describing sparks. These guides also explain how to use a chisel to size up metal composition.

I'll admit I'm the type of welder that would just as soon grab a welder and head right into the job. But Black says that knowing metal composition helps you select the right welding material, rods, or wire. And the variety of hard-facing materials is endless. For farming, where resistance to abrasion is often needed, the choices narrow. You can also get products to protect against impact, or metal-to-metal wear.

Armed with the right rod or wire, now comes the fun part: welding. But Black says that a little preparation has a big impact on the hard-facing stay-ing put. Sure, I agree, the part needs to be clean. And Black adds, "You must remove badly cracked, deformed, or work-hardened surfaces by grinding, machining, or carbon-arc gouging."

TAKE TIME TO PREHEAT

Also, be sure to preheat parts to be surfaced with a torch or electrical heater. Black explains that bringing metal up to at least a room temperature of 70°F. to 100°F. prevents "underbead cracking, welding cracking, or stress failure." Plus, preheating is the best way of slowing the cooling rate of massive or restrained parts, which are inherently crack sensitive.

Black suggests insulating the part immediately after welding with dry sand, lime, or a glass fiber blanket to minimize cooling stresses, weld cracking, and distortion.

"Never quench a weld with ice or water, as this will lead to greater internal stresses and potential weld cracking," he warns.

"And some items may require tempering or heat-treating. What this means is that you warm the item up with your torch after welding and allow it to slowly cool."

THICKER IS NOT ALWAYS BETTER

There are many more techniques to know about successfully melting hard-facing to parts. For example, more is not always better when it comes to surfacing materials. Excessive layers of hard-facing could spall in use, Black explains. And in that explanation I finally discovered what had gone wrong with that shovel I hard-surfaced years ago. That, plus I'll admit to being a lousy welder! ∎

Parts to be hard-surfaced need to be cleaned as well as preheated. Doing so prevents underbead and weld cracking and stress failures in the field.

Photograph: Doug Hetherington

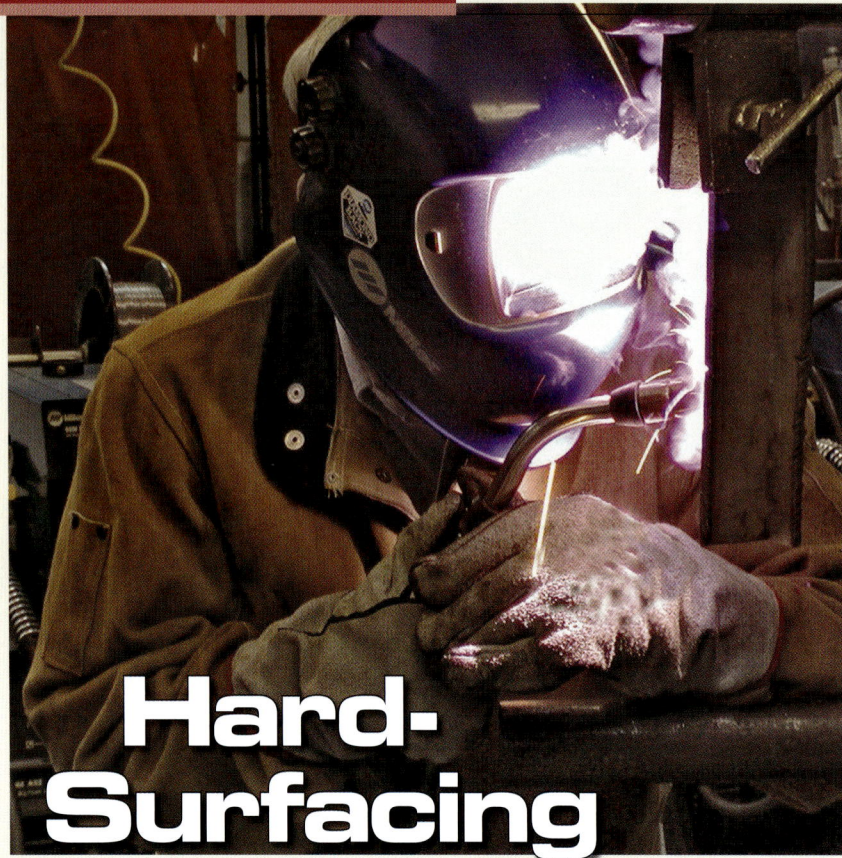

The type of welding process you use to hard-surface, such as flux-core wire welding, as well as different types of welding alloys have a huge influence on the wear or impact resistance of a processed part.

Hard-Surfacing Primer

There are as many alloys to use as there are different forms of wear

By Dave Mowitz, Machinery Editor

Hard-surfacing shovels, shanks, and drawbars seemed simple enough. Grab the stick welder, lay in welds one beside the other, and you're done.

However, success at hard-surfacing is often a 50-50 proposition. Half the time the welds stick, half the time surfaced shovels wear faster than untreated items, and most of the time the results are less than desired.

Chris Monroe says such failures aren't for lack of skill but rather lack of understanding. You may be great at joining metal but can readily screw up a hard-surfacing job, Monroe says. The cause of failures may be as simple as not selecting the right alloy.

BUILDUP AND OVERLAY

Monroe, an application technician with Hobart Brothers, explains that there are two main methods for hard-surfacing – buildup and overlay. With the buildup technique, layers of welds are used to return a part to its original dimensions. Buildup welding provides excellent impact protection but low abrasion resistance.

Overlay welding involves welding a protective layer of metal onto a new part or a used part that has been built up. Overlay surfaces provide great abrasion but poor impact resistance.

You can also use a combination of buildup and overlay welding.

"Whichever method you use, the rebuilt part is often stronger than even the original," Monroe says. "As long as the part remains sound, you can continue to use the buildup or buildup plus overlay hard-surfacing method repeatedly."

TYPES OF WELDERS

Monroe points out that stick (SMAW) welding is the most common method of hard-surfacing. "Its appeal is due, in part, to the high availability of welding alloys and to the fact that most of these electrodes lend themselves to welding on a variety of material thicknesses," he adds. "However, hard-surfacing with stick electrodes may require several layers to reach maximum wear properties, as these welding alloys tend to have a lower efficiency and a lower deposition rate."

Hard-surfacing with flux-core wire (MIG) welders provides increased deposition rates compared to stick electrodes while also improving deposition integrity, Monroe explains. "The alloys used with flux-core wire welding are generally easy to use, require minimal operator training, and can be used outdoors. Unlike stick electrodes, however, hard-surfacing with a MIG welder is best done on flat and horizontal applications and often requires two or three layers of alloy for maximum wear properties."

Photographs: Ron Van Zee

You can employ oxy fuel, gas tungsten (GTAW), or TIG welding processes for hard-surfacing. But all three processes have lower deposition rates.

DIFFERENCE IN THE BASE METAL

Generally, carbon or low-alloy steels and austenitic manganese steels are the two types of base materials that are hard-surfaced, Monroe explains. Parts containing a high carbon content have a tendency to be brittle and susceptible to cracks. Both problems increase as the carbon and alloy content of the part increases. In such cases, you may need to pre- or postheat, slow cool, or stress relieve the welds in order to ensure you have solid, long-lasting weld beads.

Preheating especially reduces the chances of developing cracks, distortion, porosity, and other weld inconsistencies. As a rule, "the higher the carbon and alloy content of a base material, the higher the preheating temperature requirement," Monroe explains.

On austenitic manganese steels, you need to take special precautions to prevent brittleness when hard-surfacing. Even though this base metal is strong and hardens under impact, preheating should not be done unless the part is less than 50°F. During welding, the base metal temperature should remain under 500°F. Austenitic manganese steels gradually become more brittle as this temperature barrier is exceeded for longer periods of time.

THE RIGHT CHOICE IN A SURFACING ALLOY

To choose an appropriate hard-surfacing alloy to protect or rebuild your parts, it's important to determine the type of wear the part will face. For example, abrasion is responsible for most wear on equip-

ment. However, there are three types of abrasion. Depending on whether your equipment is subject to low-stress scratching, high-stress grinding, or gouging abrasion, the hard-surfacing alloy you choose will vary.

Low-stress scratching abrasion – the slow wearing away of parts due to the repeated scouring – is the least severe type of abrasion. Alloys with carbide (especially chrome-carbide) are a good choice for protection. Another plus is that many carbide alloys are designed to develop stress-relieving cracks that prevent spalling, Monroe explains.

High-stress grinding abrasion is caused by materials repeatedly crushing or grinding against a part. The best hard-surfacing alloys to weld with are those containing austenitic manganese, martensitic irons, or titanium carbides.

Alloys containing carbide and supported by austenitic manganese are the best choice when encountering gouging abrasion. This type of abrasion occurs when large objects, such as rocks, are pressed against the equipment and then create gouges and grooves.

For parts that are subject to repeated impact, it is important to find an alloy with work hardening characteristics, such as those containing 11% to 20% austenitic manganese steel.

METAL-ON-METAL WEAR

Metal-to-metal wear results from the nonlubricated friction of metal parts. To prevent such wear, use a martensitic hard-surfacing alloy. Austenitic manganese or cobalt-based alloys will also work well, but it is important not to overmatch too soft of an alloy on too hard of a surface. This combination will not resist adhesive wear for long, Monroe warns.

The final consideration you need to make before choosing a hard-surfacing alloy is to determine the type

The surface left after welding is determined by the type of wear that parts are subjected to during use. The lines of hard-surface welds sometimes work best against impact wear, whereas a smooth surface is better for soil abrasion.

of surface finish you need.

For example, if you need a smooth final surface, then calculate whether the cost of grinding still makes hard-surfacing a viable cost-saving measure. Since hard-surfacing alloys range from easy to difficult in terms of grinding, you need to determine the finish prior to choosing an alloy.

"You could opt for an alloy that has slightly less wear resistance but one that provides you with the desired surface finish," Monroe explains.

"You may also consider using an alloy that can be heat-treated to soften it for machining and then brought back to the hardness necessary to protect your equipment. Likewise, if relief checks (small checks that do not weaken wear resistance) are acceptable on your surface finish, using a carbide alloy designed to be crack-sensitive may be a good option," Monroe advises. ∎

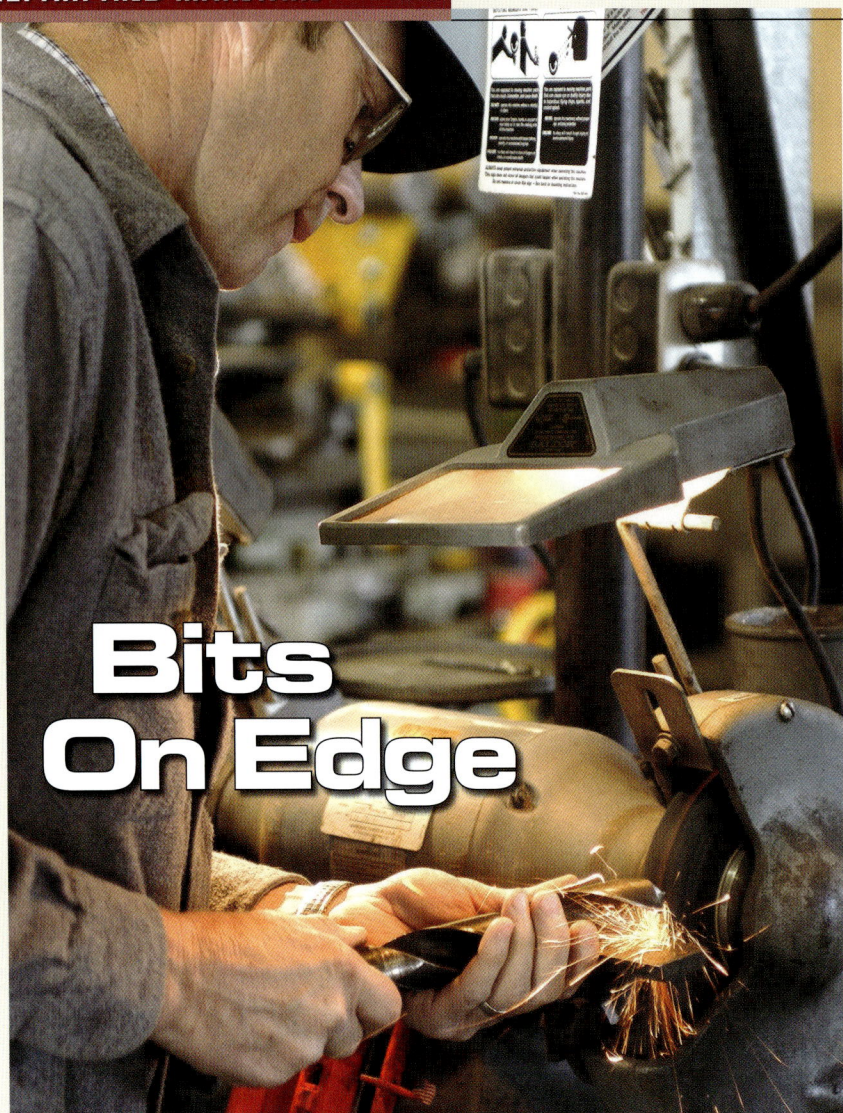

Bits On Edge

Sharpening lip angles is crucial to drill bit performance.

5 points put a sharp edge on drill bits

By Dave Mowitz, Machinery Editor

The secret to sharpening a drill bit is to understand how the bit does what it does and then to sharpen it accordingly.

One person who fully understands that secret is Dan Dovel with Drill Doctor. Dovel explains that bits do all their work at their points. And by point, he means the entire end of the bit – not just the tip in the center. The tip is called the dead center or, if so equipped, the chisel edge. The grooves running down both sides of a bit are called flutes. Their job is removing filings, so they don't need sharpening.

EXACT MEASUREMENTS

The working part of the point is the two lips, or cutting edges. The angle that these lips are cut must be kept as exact as possible.

Each lip on a high-speed steel bit is ground at a 59° angle to the center. Together the two lips form a 118° angle. To sharpen one lip at 59° and the other at 57° will have the undesired result of one lip doing all the work at half the speed.

The surface behind the lip recedes back from the lip to the heel, which is that line where the angled point meets up with the cylindrical sides of the bit.

The heel drops back from the lip to prevent the entire surface behind the lip from rubbing against its work and causing friction. That angle of heel is known as the lip clearance angle. Common lip clearance angles are 12° and should be kept to that when sharpening.

Here are five important techniques to use when grinding a bit:

1 Use a fine grinding wheel that is dressed so its surface is flat.
2 Hold the drill between your thumb and index finger. You'll want the drill tip to be exposed about 1 inch.
3 Place the back of your index finger on the tool rest with your thumbnail up and the drill at the proper angle to the wheel. The lip on the left should be visible and parallel to the grinding wheel.
4 Touch the lip to the wheel and lower the opposite end of the drill as you give it a slight clockwise twist.
5 Rotate the drill half a turn so the other lip is visible, and repeat the process described above. ∎

Clean And Lube Your Table Saw

left: After vacuuming most of the dust, blast compressed air into the saw cabinet to dislodge the remaining deposits.
below: Apply paste wax to gearing with a toothbrush.
bottom: A plastic straw delivers aerosol white lithium grease with precision.

If the moving parts of your most important shop tool don't offer peak performance, perhaps a little TLC is in order

By Dave Mowitz, Machinery Editor

If your table saw creaks and groans when you crank the elevation and blade-tilt wheels, it's long overdue for an inspection and tune-up.

Outlined here is the procedure to get your saw moving smoothly again along with important safety issues.

FIRST, CLEAN YOUR MACHINE

Uplug the saw to begin. Remove the throat plate, blade guard, and the blade. Inspect the blade for resin buildup, and clean it if necessary. Make sure the washer and blade stabilizer (if used) are clean, smooth, and flat.

Removing the drive belt and motor from the back of the saw is a fast and easy step on contractor-style models, and it dramatically improves access to the saw's interior for cleaning and lubrication. A shop vacuum with a crevice attachment will remove most of the chips, and an old paintbrush will help loosen stubborn pockets of dust.

Tilt the arbor assembly to dump more dust and use a couple of blasts of compressed air to complete the job. Make sure you've removed all dust near the stops that limit the tilt control so you'll get full travel.

If the worm gears or the rows of teeth have any residue, scrub them with a brass brush. For really tough buildup, you may have to dip the brush in paint thinner. Keep the solvent away from the arbor bearings, which are usually sealed and need no lubrication. Afterward, wipe any remaining residue from the worm gears in preparation for the next step.

TIME FOR A LUBE JOB

After all of the gearing is clean, lubricate it with a nonsilicone automotive paste wax applied with a toothbrush. Also wax the curved slots in the front and rear trunnions. Run the tilt and elevation controls through several full ranges of motion and remove all the wax, leaving only a thin film.

Push a plastic straw tip onto a spray can of white lithium grease and lubricate the pivots of the arbor assembly (where it swings upward) and the shafts behind the worm gears. This aerosol, available at auto parts stores, sprays and penetrates like a liquid and congeals into grease. Again, wipe off all the lubricant you can with a rag.

Inspect the arbor flange, making certain it's clean and smooth. Turn the arbor by hand and try to wiggle it. Any noise or sideways play indicates a problem with bearings that require immediate attention.

Blow dust out of the fence-locking mechanism. Give the fence and the entire surface of the table and extension wings a coat of nonsilicone paste wax or a special product like Boeshield T-9 (800/962-1732 or boeshield.com). ■

Photographs: Blaine Moats

Secret Life of Lubricants

The complexities of today's lubricants demand a better understanding of their needs

By Dave Mowitz, Machinery Director

My grandfather rubbed a dipstick between his fingers to determine if it was time for an oil change. My dad used to substitute engine oil for hydraulic fluid.

I'll confess to grabbing any cartridge of grease and pumping it into bearings whether they service an auger or trailer axle. After all, grease is just grease, right?

If responses to surveys posted on www.agriculture.com are a good bellwether, a great many of you think the same way. Lubricants have a tendency to be taken for granted and often misunderstood. That ignorance can lead to sloppy maintenance practices.

The problem is that today's machinery isn't as forgiving of lubrication slipups compared to equipment built even a decade ago. Take the diesels currently in use, for example. They run hotter with higher injection pressures and tighter component tolerances. Fill their sumps with SAE 30 and you void the engine's warranty. Skipping oil changes puts their multipurpose turbochargers at jeopardy, threatening a $3,000 repair job – all for the sake of saving $100 on engine oil.

LUBE MYTHS, MISTAKES, AND MISCONCEPTIONS

Misunderstandings abound when the talk turns to lubricants. Here are five common misconceptions.

■ **The W in 10W-30 represents an oil's weight.** Stede Granger of Shell Oil attributes this to confusion about the terms *weight* and *viscosity*. The W represents the winter viscosity rating of the number to its left.

■ **Lubricants have an indefinite shelf life.** Oils are long-lived. But in time, lubricants can deteriorate. Some harden like grease. Others, like engine oil, can have their additives precipitate out (solidify or gel). Eventually, most lubricants become obsolete and then fail to meet standards set forth by manufacturers. So it's recommended that you only buy the amount that you can use in a year.

■ **Automatic transmission fluid (ATF) can be added to diesel fuel to clean an engine's fuel system.** Although it is a highly refined product, ATF makes a poor detergent and using it as such can void an engine's warranty.

■ **Motor oil can be substituted for hydraulic fluid.** Although similar in structure, these two lubricants have different viscosities and additive packages. When misused, they can cause more damage than good.

■ **Different lubricants can be mixed safely.** As mentioned previously, it's a bad idea to substitute any lubricant, even different multigrade engine oils. Different brands of grease, for example, can be incompatible. ■

Photographs: Ron Van Zee

And it's not just changes to engine technology that are pushing the limits of lubricants. Oil chemistry itself has undergone radical changes of late, as seen with the latest diesel oil grade, CJ-4, which uses significantly less phosphorus. Starting next year, you'll only be allowed to burn ultra-low sulfur diesel that uses a fraction of the sulfur (an additive with lubricating abilities) compared to today's fuel.

Phosphorus and sulfur? Sounds more like a fertilizer than a lubricant. What happened to plain oil?

Oil hasn't been just oil for a long time. Today's lubricants are a virtual chemistry set. They comprise packages of additives – like sulfur and phosphorus – and dozens of other

Antiwear additives are king in hydraulic oil formulations. But "in tractor hydraulic fluid, it is critical that your additives are balanced so your equipment can perform all functions without compromise," advises Sheri Barta of John Deere.

ADDITIVE PACKAGES PINPOINT PROTECTION

Even the highest quality base oil needs the extra oomph to meet manufacturer lubrication requirements. This is where additives come into play. The balancing act of creating an additive package for a particular need, however, is stunningly complex. Too much of a particular additive, such as a dispersant, could compromise the effectiveness of an antiwear agent.

Also, some additives find common use in different lubricants. Such is the case with zinc dithiophosphate (ZDDP) found in many lubricants like engine and hydraulic oils. But the amount of ZDDP in hydraulic oil is different than in engine oil, as pumps need to provide more metal-to-metal contact protection.

For this important reason, engine oils should never be used in hydraulic systems, warns Brendan Casey of www.InsiderSecretsTo Hydraulics.com.

EXAMPLES OF COMMON OIL ADDITIVES

Additives	Functions
Detergents, dispersants (calcium, magnesium, metallic soaps)	Remove soot, carbon, varnishes, and unburned fuel
Antiwear agents (zinc dithiophosphate)	Minimize metal wear
Pour point depressants	Allow oil to flow in cold temperatures
Antifoam agents	Retard foaming by removing bubbles
Antioxidants (zinc dithiophosphate)	Slow deterioration of oil from high temperatures
Corrosion inhibitors (zinc dithiophosphate)	Prevent rusting, reduce corrosion in engine during storage
Emulsifiers	Suspend and remove moisture
Extreme pressures (sulphur-phosphorus, chlorine)	Minimize metal-to-metal contact

Furthermore, be wary of oil performance-enhancing products. Oil companies have already gone to great lengths to create additive packages to meet specific vehicle operating requirements set forth by manufacturers, explains Ed Hackett of the University of Nevada.

"If you add anything to oil, you may upset that lubricant's additive balance, and prevent the oil from performing to specifications," he warns. ■

Automatic transmission fluid (ATF) may be called a fluid, but it is every bit a mineral-based oil. The confusion comes with its color (often red or green), which merely differentiates specification sources (like Dexron or Mercon). Some manufacturers recommend different ATF products for different vehicle uses such as frequent trailer towing by heavy-duty pickups. ATFs employ a battery of additives crucial to their performance led by viscosity improvers that boost the lubricants' operational temperature range.

compounds that add extra oomph. This allows the oil to work harder and longer in a wider range of conditions, explains Mark Bitner of CITGO.

OIL CARRIES THE WORKLOAD

Still, the foundation of any lubricant is its base oil. Diesel oils are 75% to 85% base oil, with the remainder of their volume consisting of chemical compounds. That percentage is higher in some lubricants (like gear lube) and lower in others (like transmission fluid).

You have likely noticed a wider variety of base oils being offered today, ranging from mineral-base fluids to synthetic oils and combinations of both. Regardless, they all come from

SHORT COURSE ON ENGINE OIL TERMS

As lubricant formulations have become more complex, so has the vocabulary used to describe them. Following are some basic terms you should know when buying engine oil. Find a library of lubricant language on the Web at www.agriculture.com.

API: The American Petroleum Institute establishes all motor oil classifications in conjunction with engine manufacturers. If an oil doesn't carry an API rating, don't buy it.

Back serviceable or compatible: This term designates an oil's ability to provide protection when in use on all engines, both old and new.

EP: This term designates a lubricant's ability to prevent sliding metal surfaces from seizing under extreme pressure conditions.

Flash point: This is the lowest temperature at which vaporized oil will flash when exposed to a flame.

Lubricity: This is the ability of oil to stay in place to prevent wear.

Pour point: An indicator of the ability of a lubricant to flow in cold temperatures, this term is interchangeable with cloud point. A pour point depressant is an additive used to lower the temperature fluidity of an oil.

SAPS: This is an abbreviation for sulfated ash, phosphorus, and sulfur. The EPA has greatly restricted SAPS content in engine oils.

Total base number (TBN): This is a measure of an oil's ability to neutralize acids formed during engine operation. TBN is sometimes referred to as Base Number.

Shear: This term describes the ability of oil to remain stable, intact, and in function when under load.

Sulfated ash: This is the noncombustible residue of a lubricating oil and fuel. ∎

the same base stock: refined petroleum. There is this misunderstanding that synthetic oils are fluids separate in nature from mineral oil, observes Reginald Dias of ConocoPhillips Lubricants. Synthetics are created through a chemical reaction process. Mineral oils are obtained via traditional distillation methods.

SYNTHETIC ADVANTAGES

The reaction process used to produce synthetics, called polymerization of hydrocarbons, produces oils that are more uniform molecularly. That uniformity imparts performance advantages such as reduced oil consumption, longer drain intervals, and improved high-temperature stability.

The capacity of synthetics to operate in a wide range of temperatures without breaking down explains why we see more synthetic versions of gear lube, hydraulic fluids, and grease. These offerings have come in response to manufacturers pushing the performance of mechanical components on their machines.

But synthetics come at a higher price, which is a matter you must debate with your checkbook and supplier. When in doubt, refer to the specifications listed by your equipment's manufacturer. "That must be the starting point when shopping for any lubricant," says CITGO's Betner. "We formulate our lubricants to meet their specifications."

IT'S ALL ABOUT VISCOSITY

No matter if an oil is mineral-base or synthetic, its single-most important performance characteristic is viscosity. But viscosity may be the most misunderstood term in agriculture. Viscosity is the measure of an oil's resistance to flow (shear stress) under certain conditions. To simplify, viscosity is the tendency of oil to stay put when pushed (sheared) by moving mechanical components. As such, the thicker the base oil,

Designer colors for grease? Color only differentiates grease brands or grades, which are far more complex than we often realize. The base oil in grease can come in a wide range of viscosities that affect performance. For example, you don't want to use grease with a heavy base oil in universal joints, as its oils must easily flow in between needle bearings. But heavy base oils are ideal in slow-speed and high-load applications. Then, too, the soap (or thickener) used to suspend oil can be formulated from a variety of substances from lithium (the most common soap) to complexes of calcium, barium, or aluminum. These thickeners play a crucial role in releasing oil. "Thickener acts like a spider's web," says Mark Betner of CITGO. "It interlaces and locks in the oil until it's needed for lubrication."

the higher its viscosity rating. That rating is represented by all those numbers printed on an oil can.

There are two viewpoints of oil's resistance to flow that the machine designers follow. One is the measure of how the fluid behaves under pressure, such as in a pressurized hydraulic line. This property is called absolute viscosity (also known as dynamic viscosity), and this is measured in centipoises (cP).

The other measure is how the fluid behaves only under the force of gravity. This is called centistokes, the measure used to describe engine oils.

But why the need for two numbers? An oil's viscosity changes as its operating temperature is raised or lowered. And this characteristic has a huge effect on its ability to provide protection.

RANGING VISCOSITY INDEXES

The ability of an oil to perform at various temperatures is represented by its viscosity index, or VI. This number represents the degree of change in viscosity within a given temperature range. A high VI indicates relatively small change in viscosity with temperature change. A low VI reflects a larger viscosity change with temperature.

Achieving high VI ratings required by motor oils, for example, was accomplished by including additives called viscosity modifiers or VI improvers in the base oil. VI improvers are high-molecular weight polymers that remain inert at low temperatures. As oil is heated, they expand to help the oil maintain its viscosity.

VI improvers create the multigrade

lubricants that are the mainstay of motor oils today. For example, a 10W-30 oil behaves as 10-grade oil at low temperatures but gives the protection of 30-grade oil at the high operating temperatures. Multigrade oils can have a VI well over 100; single-viscosity oils have a VI of about 100 or less.

"No single-grade oil has passed American Petroleum Institute standards since 1991," says Stede Granger of Shell Lubricants.

Here is one last differentiation that is crucial for motor oil use.

Lubricants designated with a C such as CJ-4 are for use in diesels. Oils with an S designation are for use only in gasoline engines. You can use a C-rated oil in a gas engine since diesel oils offer superior protection. But never pour a gas oil in a diesel, or a potential early overhaul is in that engine's future. ∎

FOUR MOST LETHAL OIL CONTAMINANTS

As well crafted as modern oils are, their capability to protect engines can be negated by contaminants. The following four contaminants are the worst in that regard. By themselves, they can cause oil to break down prematurely. But in combinations, these contaminants are lethal to engines.

1. Water: The most destructive of all contaminants, water interferes with oil film protection, attacks additives, boosts oxidation, and increases the corrosiveness of common acids. Emulsifiers can mop up some engine condensation, but too much moisture overwhelms oil and creates globs

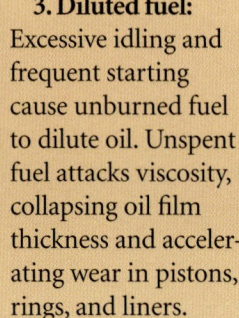

Excessive soot accumulation jeopardizes piston rings causing increase blow-by of fuel.

of sludge that restrict its flow.

2. Glycol: The glycol portion of antifreeze boosts metal wear rates (10 times greater than water contamination) while reducing oil's viscosity.

3. Diluted fuel: Excessive idling and frequent starting cause unburned fuel to dilute oil. Unspent fuel attacks viscosity, collapsing oil film thickness and accelerating wear in pistons, rings, and liners.

4. Soot: All diesel engines generate soot. But soot concentration polishes off the antiwear films on boundary zones (such as cam and cam followers), and it jams up rings (called carbon jacking). ∎

LOW-SULFUR DIESEL MANDATED NEXT YEAR

If you are driving a 2007 or younger diesel pickup, you already know it can only burn ultra-low sulfur diesel fuel (ULSD) and use low-sulfur oil (CJ-4). To do otherwise voids that vehicle's warranty protection. In fact, starting in 2007 all diesel-powered vehicles on the highway had to switch to ULSD.

Now the rule is coming to agriculture. Starting in 2010, all diesels operating in a pickup, semi, tractor, or combine must burn low-sulfur fuel.

This presents a bad news/good news situation. The bad news is that ULSD fuel and CJ-4 oils are more expensive. Also, ULSD has lower cetane levels so it generates slightly less power. And there are some fears that ULSD could cause

seals to leak on some older diesels. The good news is that ULSD fuel will be plentiful thanks to the fact that refiners have already had to meet the supply needs of over-the-road vehicles. Another positive note is that CJ-4 oils are fully backward serviceable and compatible.

"There is no need to buy another oil for older engines. Besides, CJ-4 provides wear protection and contamination control superior to older oil grades," says Stede Granger of Shell Oil.

The best bet is to use up all existing CI-4 Plus oil (if you haven't already) and stock strictly CJ-4 from here on out. "Consult with engine manufacturers about drain levels on older engines when using CJ-4," says Mark Betner of CITGO. ∎

A sign of the coming times in farming can be found on diesel pumps at filling stations. All off-road vehicles must burn ULSD starting in 2010.

Coolant Blues – or Greens

left: Cavitation erosion is at fault in pitting wet-sleeve liners in the engine. Eventually these pits perforate liners and allow coolant to leak into an engine's crankcase.

above: Corrosion has eaten away the metal of the engine tubing.

40% of diesel engine downtime can be attributed to coolant problems

By Dave Mowitz, Machinery Editor

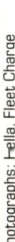

Photographs: Hella, Fleet Charge

Coolant gets no respect. The green stuff circulating through radiators is easily ignored since it often keeps working long beyond its prime to cool engines.

Yet coolants do more than cool; they must protect engine parts in the process. Failing to do so causes approximately 40% of diesel engine downtime, engineers estimate. "A fresh fill of antifreeze is good insurance against trouble and costly repairs," says Utah State University Extension engineer Von Jarrett.

Major problems relating to cooling breakdowns include metal corrosion, cavitation, scale deposits, and inhibitor dropout or green goo.

CORROSION, CAVITATION KILLERS

Corrosion is the natural tendency of metals to revert to their ore form. "All metals exposed to coolant will corrode," says Frank Cook of Old World Industries, a coolant manufacturer. "Proper coolant chemistry slows the rate of corrosion by forming a protective layer on metal surfaces," he says.

Borate, for example, is an inhibitor that maintains pH between 8.5 and 11, providing protection for iron and steel components. Silicate and nitrite in coolant protect aluminum and ferrous metals like iron and steel.

Nitrite actually does double duty, protecting wet-sleeve cylinder liners against cavitation. This form of metal erosion occurs during an engine's power stroke, which causes the outside wall of a liner to move away from the coolant and create, for an instant, a near vacuum. This low-pressure pulse causes the surrounding coolant to boil and to form tiny bubbles.

The liner then returns to its position at high velocity and presses against the bubbles. The bubbles implode (collapse) against the liner wall surface at pressures up to 60,000 psi and blast small pits in the steel. Nitrite forms a thin protective film on the liner wall and acts as a barrier against pitting.

Unlike cavitation, scale adds to liners and other surfaces in an engine's cooling system. Just 1/16 inch of scale reduces heat transfer efficiency by 40%. Since scale tends to form in specific areas of the hot side of the engine, it causes localized hot spots and distortion. The harder the water in a coolant, the greater the amount of scale formation. This is why deionized water is used in coolants.

Too high a concentration of two common antifreeze inhibitors, phosphate and silicate, can also cause overheating by coating coolant system surfaces with green goo. This situation, called inhibitor dropout, occurs in a lot of failed water pumps, thermostats, radiators, and heater cores.

The key to avoiding green goo is to use a "high-quality, fully formulated precharged coolant," says Craig Gullett of Fleet Charge, an automotive supplier. He urges using a fully formulated diesel coolant that meets The Truck Maintenance Council Recommended Practice RP-329. Look for this designation on the coolant label. ∎

Does Coolant Color Matter?

Color more often identifies manufacturers' differences

By Dave Mowitz, Machinery Editor

Wait for a radiator to completely cool before checking fluids. Use a bulb hydrometer to measure the concentration of antifreeze. To determine fluid quality, use test strips like FleetGuard's Quik-Chek strips (above).

Time was when coolants were one color – green. Then came extended-life antifreeze. That's when marketers started selling their product in designer colors. General Motors brought out Dex-Cool in a distinctive orange color. Caterpillar engines now come with red-orange coolant. Other engines offer blue, purple, and even pink antifreeze. So does color make a difference in quality?

Not really. The glycol base and additive package in coolants are generally clear. Dye is added to help differentiate coolant type (regular or extended life) or a manufacturer's distinctive additive package.

Generally, green coolants are for light-duty use and don't have extended-life characteristics (although some Japanese cars use an extended-life antifreeze that is dark green).

Other colored coolants generally signify they have extended-life characteristics. However, color doesn't guarantee extended-life quality. So the best advice when buying antifreeze is to pay less attention to color and just read its label to determine which vehicles it can be used in.

BACK WHEN IT WAS SIMPLER

"Way back when engines were simpler, the primary goal of radiator fluids was to prevent engines from freezing," says Carmen Ulabarro, a coolant specialist with Chevron. "Today, however, coolants are crucial for heat transfer and corrosion protection as well."

This explains why radiator solutions no longer employ methanol, which worked to prevent water from freezing. But come summer, methanol tended to boil over.

To solve this problem, the automotive industry switched to glycol as the base for coolant. Glycol works great as an antifreeze. But it isn't very good at conducting heat from the engine to the radiator.

THE NEED FOR WATER

Simple water, on the other hand, excels at this job. Which explains why you shouldn't use pure coolant in a radiator. This practice can cause an engine to run at the wrong temperature. Water also works to activate the chemicals that protect against rust and corrosion.

The glycol in antifreeze is either ethylene glycol (EG) or propylene glycol (PG). What are the differences between the two? EG, which is the most commonly used coolant base, is more toxic than PG. Otherwise, the differences are nominal.

Whether a coolant has an EG or a PG base, what really separates coolants is their chemical composition.

Coolants commonly used in older engines use inorganic acid technology (IAT) that provides a life span of two years or 30,000 miles. IAT coolants are typically green in color.

Coolants with organic acid technology (OAT) composition, introduced in 1994, have an extended life span of five years or 150,000 miles. OAT coolants come in a variety of colors.

Today there are hybrid combinations of both IAT and OAT compositions.

COOLANT CLASSIFICATIONS

Coolants are often classified by their type of use – automotive, heavy-duty, and universal (for both car and heavy-duty engines). Differences between these classifications are often determined by the level of additives they contain to protect against rust and corrosion.

For antique tractor engines, use coolants rated for light-duty or universal use. When it comes to high-power modern tractors as well as trucks, go with the manufacturer's coolant recommendation. ∎

Photograph: Baldwin Filters

Safer, Cheaper Soy Solvents

New solvents are less toxic and more environmentally friendly

By Dave Mowitz, Machinery Editor

A ready replacement for more hazardous and expensive parts washer solvents can be found in the soybean-based methyl soyate.

Cleaning parts just got a lot safer. And you can thank farmers for that.

Until recently, your choices of parts washer solvents included products like trichloroethylene (TCE), perchloroethylene (Perc), methylene chloride (MeCL), mineral spirits, and toluene. Beyond the fact that some of these solvents are darned difficult to pronounce, they have the additional drawbacks of being hazardous to health and environmentally unfriendly.

A number of alternatives have come on the market of late like aqueous solutions, d-limonene and N-methyl-pyrrolidone (NMP). Yet even some of these solvents are not environmentally friendly, can be expensive and, well, are still hard to pronounce.

Leave it to farmers to find a solution with a simple name! Aided by funding from the United Soybean Board (USB), chemists have developed a solvent from soybean oil that is not only biodegradable, but is also less toxic than conventional solvents.

SLOW TO CATCH FIRE

Another advantage of methyl soyate is that it has a high flash point, which means it won't spontaneously combust until it reaches a temperature of 360°F.

But the best thing about soy solvents is their price. Current pricing for methyl soyate ranges from 55¢ to 65¢ a pound, according to the USB. This compares to 65¢ a pound for mineral spirits, $1.15 a pound for d-limonene, $1.80 a pound for NMP, or $10 a gallon for aqueous cleaners.

"And the escalating prices of petrochemicals have caused many industrial solvent prices to rise beyond methyl soyate," says Chris Toebben of the USB.

That organization estimates that more than 10 manufacturers now offer soy solvents. For example, a 5-gallon container of Bean-e-doo Parts Washer solvent sells for $69.95. Bean-e-doo is sold by Franmar Chemical (800/538-5069 or www.franmar.com).

You can obtain a list of other companies selling methyl soyate products by calling 800/989-8721 or going to www.unitedsoybean.org. ∎

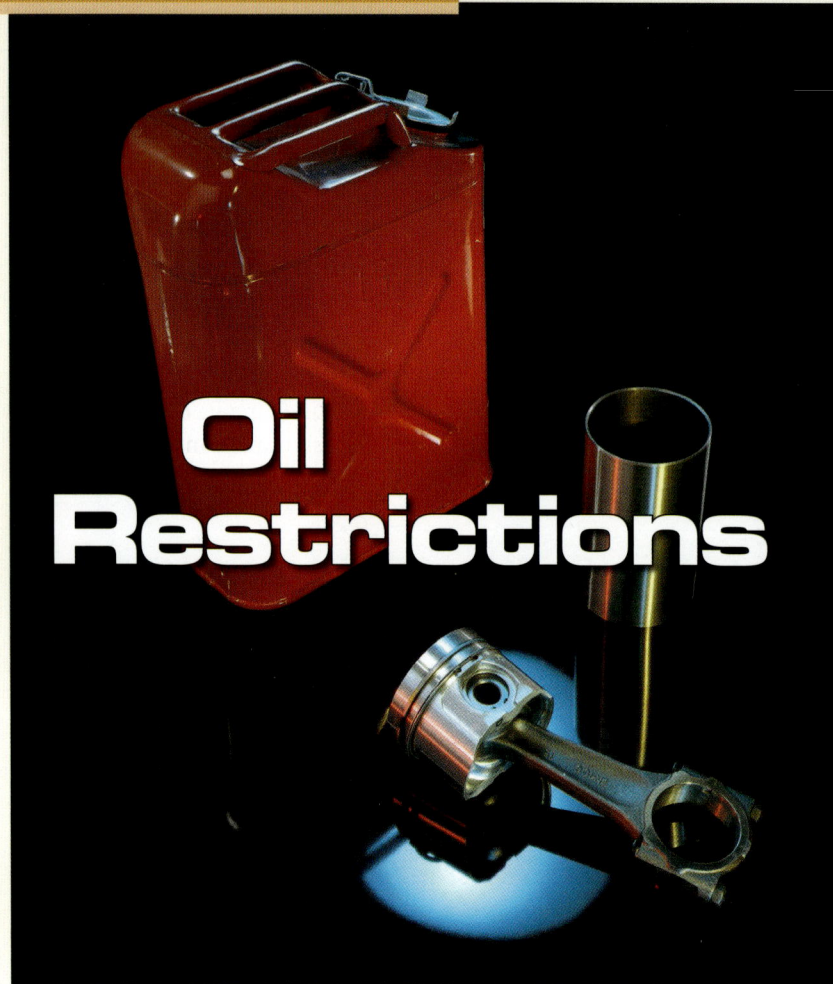

Oil Restrictions

The new generation of CJ-4 diesel engine oils are 10% to 15% more expensive than earlier grades but have superior wear protection and ability to keep contaminants suspended.

Latest grade of oil becoming mandatory on ag engines

By Dave Mowitz, Machinery Editor

The mandate that all on-highway vehicles use only ultra low-sulfur diesel fuel (the switch for off-road machinery comes in 2010) has spawned an entirely new class of diesel engine oils.

CJ-4 oil represents one of the most significant advances in lubricants in decades, observes Stede Granger of Shell Lubricants (www.shelllubricants.com). "This specification of diesel oil provides superior wear engine protection, higher temperature stability, and ability to control contaminants from affecting parts."

ADDITIVE RESTRICTIONS

What sets CJ-4 oils off in a class of their own is that the grade represents the first time that restrictions have been placed on the oil additives. For example, CJ-4 oils cannot contain more than 0.4% sulfur, 0.12% phosphorus, or 1% ash. In comparison, the old oil grade of CI-4 contained 1.3% to 1.5% ash. This particular additive is used in combination with phosphorus to enhance a lubricant's ability to provide protection against oxidation. Replacing ash and phosphorus are such additives as zinc, which provides superior wear protection.

Additive restrictions came as a result of the low-sulfur fuel mandate as well as increasing use of post-combustion treatments, such as catalytic filters, in on-highway engines to reduce pollution. Similar technology will be employed in ag engines in the near future.

Modern diesels also use technology such as exhaust-gas recirculation (EGR) to cut pollution.

HOTTER RUNNING ENGINES

Engines equipped with EGR run hotter and require oil with improved oxidation resistance to prevent breakdown. "CJ-4 oil provides temperature stability, which greatly reduces engine wear," explains Nicole Fujishige of Chevron Products (www.chevron.com). "Although backwards compatible (can be used in older engines and mixed with CI-4 oil), CJ-4 oils are required in all on-highway engines."

Many lubricant companies have stopped making the old CI-4 oil. This explains a recent move by suppliers to switch farms to CJ-4. "CJ-4 is being required in more tractor engines now and will be mandated in the future," Granger says. "Rather than risk voiding your engine's warranty, we recommend switching entirely over to CJ-4, while using up CI-4 oils only in older engines until supplies are gone." ∎

Photograph: Baldwin Filters

Media Matters

Photograph: Ron Van Zee

Not all oil filters are built the same, so follow the manufacturers' recommendations on your engines.

Not all oil filters are created the same, and this has an effect on their performance in the field

By Dave Mowitz, Machinery Editor

There is a good reason engine manufacturers ask you to use filters that meet their specifications. Not all oil filters are built the same, they warn. And quality differences among filters could amount to an early engine overhaul.

Many engine oil filters, for example, use pleated cellulose or cellulose-blended media that are adequate to remove larger particles in oil. As the size of contamination gets smaller, filter manufacturers will go with a complex media created from blending cellulose and synthetic fibers.

And to remove extremely small particles, such as when filtering hydraulic oil, filter media are made exclusively from synthetic microfibers. Such man-made glass fibers have a more uniform media opening size, which provides better flow performance. Plus, they can hold more dirt per square inch of media than cellulose blends.

MICRON SIZE MATTERS

That opening size is expressed in micron size, explains the Filter Manufacturers Council, a group which represents a wide variety of companies. A micron is another term for micrometer, which represents 1 millionth of a meter. For comparison sake, the diameter of a human hair is 70 microns.

Micron rating is a generalized way of indicating a filter's ability to remove contaminants. The assumption is that 15 micron media is good at removing particles 15 microns in size and larger.

But a low micron rating isn't a complete indication of filter quality in terms of efficiency. Until recently there wasn't a universally accepted method to measure the size particles a filter's media could capture and hold, explains Robin Bumpass of the Council. Depending on the test method used, media could be rated with different micron ratings.

REFER TO THE MULTIPASS TEST

That changed with the creation of the multipass test. Recognized by the Society of Automotive Engineers, this evaluation provides a percentage describing how efficient a filter's media is at removing specific particle sizes. The multipass test can determine the total contaminant holding capacity of a filter as well as some of its differential pressure capabilities, Bumpass says.

When comparing filter make, use the multipass rating as a guide. But also check that the filter meets manufacturer specifications, Bumpass advises. Those specs provide for a balance between efficiency and capacity in a filter.

For example, a very high-efficiency filter may have lower dirt-holding capacity to the point that, in certain situations, it will cut short its life. This partially explains the difference between full-flow and bypass filters on an engine. A full-flow filter consumes approximately 90% of the oil flowing through an engine.

On the other hand, a bypass filter, which is typically high efficiency in design, consumes less oil. This might be just 10% of the oil coming out of the full-flow filter.

After passing through the bypass filter, that oil is returned to the engine's sump. This high-efficiency filtration removes fine contaminants not caught by the full-flow filter while not compromising total oil flow. ∎

Hungry Air Tools

Air requirements for impact wrenches – the most-used tool in farm shops – vary widely by the size of the tool. For example, an impact wrench with a 3/8-in. shaft consumes about 3 cfm. Pick up an impact wrench with a 1¼-inch shaft (which is getting more popular on farms) and air demand shoots up over six times to as much as 19½ cfm. Operating another tool at the same time can quickly starve one or both tools.

The growth in air tools is taxing some compressors

By Dave Mowitz, Machinery Director

When pneumatic tools first got popular with farmers, old compressors were likely adequate to run a couple of impact wrenches and still allow a tire to be aired up.

But as farmers discovered the power and convenience of air tools, their fleet of pneumatic gadgets started to grow in a big way. The problem is that an old compressor may not be capable of expanding its output in airflow and pressure to keep up with additional air tools.

"An air-starved tool won't operate at its peak power," warns Joel Stevens of Chicago Pneumatic (www.chicagopneumatic.com).

CALCULATING AIRFLOW NEEDS

Determining whether your old compressor is able to keep up with that new fleet of air tools and calculating what size new compressor you'll need to purchase are relatively simple processes. Start by discovering the requirement of your air tools, according to guides provided by Devair Inc. (www.devair-compressors.com), Ingersoll Rand (www.products.ingersollrand.com), and Northern Tool & Equipment (www.northerntool.com).

Most tools will list both their required air pressure (in psi) and airflow (in cfm). For example, a typical impact wrench equipped with a ¾-inch shaft requires 75 psi at 8 cfm. Check the tool's owner's manual if this information is not printed on the tool itself. If the manual is unavailable, use the tool chart listed on the next page.

The next step is to start adding up the cfm requirements of all the tools that could be used simultaneously. This represents the minimum cfm a compressor must have to supply your shop.

INTERMITTENT VS. CONTINUOUS TOOLS

Keep in mind that tools are classified as either intermittent or continuous. The vast majority of tools in use in a farm shop are intermittent in nature. For example, hand-triggered tools such as impact wrenches or grinders are intermittent because they are only operated for short periods of time.

This contrasts with continuous tools such as air motors or sand blasters that run nonstop for a long period of time, points out Chicago Pneumatic's Stevens.

Next, add the cfm for each tool on the intermittent list. Now multiply that total by a factor of .40 to reflect an approximate 40% intermittent use of the tools.

As for continuous tools, use their full cfm value in your calculations.

Now, add the intermittent and continuous tool totals together. Take that figure and divide it by .75 to factor in downtime for the compressor to stop and cool down between recharging cycles.

This final figure represents the minimum air supply a compressor must deliver to service all pneumatic shop tools.

Horizontal grinders are the air hogs of hand tools, as they consume huge amounts of cfm to maintain high rpm under load. Their growing popularity in farm shops has led to existing compressors failing to keep up with shop air needs.

PNEUMATIC TOOL AIR REQUIREMENTS

Tool	Pressure Range (PSI)	CFM Consumed
Air filter cleaner	70-100	3
Air hammer	90-100	4
Bead breaker	125-150	12
Blow gun	90-100	2½ to 4
Burring tool	90-100	4-5
Cylinder hoist	70-100	1⅓
Die grinder	70-90	5
Drill, ⅛" to ⅜"	70-90	4
Drill, ⅜" to ⅝"	70-100	7
Floor jack	125-150	6
Grease gun	120-150	3
Grinder (small)	70-100	5⅓
Grinder (medium)	70-100	8½
Horizontal grinder (4"-6" dia.)	70-100	21
Horizontal grinder (8" dia.)	70-100	28
Hammer, chipper	70-100	7
Hammer, scaler	70-100	4
Hydraulic lift*	145-175	5¼
Impact wrench (⅜" shaft)	70-90	3
Impact wrench (⅝" shaft)	70-90	5
Impact wrench (¾" shaft)	70-90	8
Impact wrench (1" shaft)	70-90	12
Impact wrench (1¼" shaft)	70-100	19¼
Pneumatic jack	120-150	½
Ratchet ½"	70-100	5
Riveter	70-100	1
Sandblaster (⅛" nozzle)	80	15
Sandblaster (³⁄₁₆" nozzle)	80	40
Sander (5" dia.)	70-100	4-6
Sander (7" dia.)	70-100	10
Screwdriver, #6 (to ⁵⁄₁₆" screws)	70-100	8½
Spray gun	75-100	5 to 8
Tire changer	125-150	1
Tire hammer	90-100	12
Tire inflation line	125-150	1½ to 2

* For an 8,000-lb. capacity hydraulic lift, add 0.65 CFM for each additional 1,000 lbs. of capacity.

FIGURING PRESSURE NEEDS

You also need to factor in psi requirements of tools to assure that a compressor can supply needed pressure. There is no need to perform any detailed calculations to estimate psi needs, however. Instead, find out what pneumatic device in your arsenal requires the most pressure to operate. Bump that single figure up slightly to compensate for compressor operating irregularities.

For example, if you have determined that the tool you use with the highest-listed pressure requirement is a pneumatic jack (which operates at 130 psi), then that is the minimum pressure your compressor must deliver to the jack. So to be safe, figure on needing a compressor that delivers 135 to 140 psi.

Don't overdo it on pressure, however. Running a compressor at 150 psi to feed a jack that requires 130 psi just wastes electricity and can wear out a compressor prematurely.

COMPRESSOR OUTPUT DIFFERENCES

When your total cfm and highest psi requirements are determined, compare those figures to the output of the compressor you own or are going to buy. Look to the operator's manual of existing compressors for these output figures. Otherwise, ask a salesperson to confirm the capabilities of new compressors.

If you are shopping for a new compressor, be warned that there is a difference between displaced cfm and delivered cfm.

Delivered cfm is sometimes referred to as free air.

Displaced cfm is the airflow produced by a compressor working in a perfect environment at 100% efficiency. This rating can be misleading, as no compressor was ever made to run at 100% efficiency. Therefore, stay strictly to delivered cfm when sizing up a compressor.

You may also notice that manufacturers will list output in acfm (actual cubic feet per minute) or icfm (inlet cubic feet per minute). But don't fret over these differences because both measurements are essentially the same.

DIFFERENCES AMONG TOOLS

The other consideration when sizing up compressed needs is to take into consideration the huge differences between air tools.

With an impact wrench, for example, air consumption goes up exponentially with wrench size, but pressure remains relatively the same across this range of wrenches.

Another example is a hydraulic lift. Its airflow needs are low at 5¼ cfm. But its pressure requirements are huge, ranging from 145 to 175 psi.

Other devices that fall in the category include pneumatic grease guns, jacks, inflation lines, and tire changers. ∎

Air Impact Wrenches

Tools deliver high torque output with little effort

By Laurie Potter, New Products Editor

Impact wrenches are one of the most commonly used pneumatic (also referred to as air) tools. These versatile tools can produce thousands of foot-pounds of torque.

With plenty of nuts and bolts on farm equipment to be loosened and tightened, an air impact wrench that can tackle the task with plenty of torque is a must.

A socket wrench designed to deliver high torque output with minimal exertion by the user, the impact wrench stores energy in a rotating mass, then delivers it suddenly to the output shaft. This power makes it one of the most commonly used pneumatic tools.

Impact wrenches are available in a wide range of socket wrench drive sizes (from a ¼-inch square drive to 3½-inch and larger square drives) and in several styles, including inline, butterfly, and pistol grip.

With the inline style, the tool is held like a screwdriver with the output on the end. A butterfly style is a special inline form. It has a large, flat throttle paddle on the side of the tool that can be tilted to one side or the other to control the direction of rotation rather than requiring a separate reversing control. The pistol grip user holds a handle, which is at right angles to the output.

WHAT TO LOOK FOR

Nick DeSimone, air tools product manager for Porter-Cable and Delta machinery, says there are four things to consider before purchasing an impact wrench: clutch, retainer, power, and ergonomics.

1. Clutch. "The two most common mechanisms are the pin clutch and the twin-hammer design. The main difference between the two is that the pin clutch takes longer to arrive at the maximum torque. For example, the longer the trigger is depressed, the higher the torque will go until you reach the maximum torque of the tool," he says.

2. Socket retainer. The two most popular retainers are the pin and friction ring (also known as hog ring). "The pin retainer has the most positive engagement," says DeSimone. "The retention of the pin design can be maximized by aligning the pin to the hole that is found in the side of the socket where it mates to the square drive of the tool. It ensures the socket doesn't come loose from the tool." The friction ring design is convenient for users who make many socket changes.

3. Power. The amount of power is measured in foot-pounds of torque. Depending on bolt size and grade, this will determine how much torque you need and which impact wrench will match your application. "Most general applications can be handled with a range of 400 to 500 foot-pounds of torque," says DeSimone.

"An impact wrench on an application uses about 60% of its torque capacity," says Larry Outtrim, national sales manager for RediPower. "This is because most impact wrenches reach 60% of maximum torque in about three to five seconds, which is the time to target for most applications."

4. Ergonomics. "These tools operate on a principal of impact to generate force. Oftentimes high vibration can be experienced," DeSimone says. Tools with an ergonomic padded grip can lessen these effects.

ADVANTAGES OF AIR

One of the big advantages of air tools is that they don't require their own motors. An air compressor motor converts electrical energy into kinetic energy. In addition, air tools have a higher power-to-weight ratio, which means you get a smaller, lighter tool to tackle the same job.

"An impact wrench is a versatile tool that can produce thousands of foot-pounds of torque. Simple maintenance will allow these tools to last for decades," Outtrim says.

Depending on features, you can pay anywhere from $90 to more than $400 for a smaller wrench. ∎

Air Compressors

A T30 air compressor with a vertical tank from Ingersoll-Rand provides plenty of power for tightening lugs.

Farmers share tips on what to look for when buying stationary compressors for the shop

By Lisa Foust Prater, Commerce Editor

Farmers have plenty to say about air compressors. Here are some of the responses when the Shop Talk discussion group at Agriculture Online® (agriculture.com) was asked, "What features would you look for in the ideal farm shop air compressor?"

"First, decide what you need to flow for cfm (cubic feet per minute, which measures the volume of air delivered)," Blake Keithley tells the group. "Then, decide if you need a single or two-stage. From there, decide on the pain you can bear pricewise."

Many farmers agree that cfm is the best way to rate a compressor. "I'd recommend at least 15 cfm. You won't regret having a big compressor when you start running that sander or impact," posts Downwardspiral.

HOW MANY HORSES – REALLY?

Horsepower is key, the farmers say (5 hp. or more is recommended), but it can be tricky to determine. "Watch for models with the hp. rating on a decal on the tank. Always look at the hp. rating on the motor," Vinnie says. "You may want to check the amp draw with a meter to see the load at shut-off. Check the duty cycle. I've seen 6-hp. decals on 2-hp. machines."

Indiana Jim uses a few simple formulas to avoid confusion with hp. ratings. "Volts × amps = watts," he says, and "watts / 746 = true hp."

Since most farmers use their compressors more often and with more demanding tools than the general do-it-yourself population, they recommend two-stage compressors. Single-stage models are generally used for pressures between 70 and 90 psi, with a maximum of 135. Two-stage compressors are used for pressures above 100 psi, up to 175.

When it comes to the tank, the group says it should be at least 60 – preferably 80 – gallons in size. And while horizontal tanks have their benefits like ease of servicing, space-saving vertical tanks are favored. Cast iron pumps are also preferred. The farmers recommend avoiding oilless compressors and aluminum units.

Bringing all these desired features together into one unit doesn't come cheap, but most Shop Talk participants say it's worth the money. ∎

CFM FOR AIR TOOLS

Here are some common air tools and their average cfm at 90 psi, based on a 15-second run, from northerntool.com. For tools with a higher cfm, using a larger tank will save wear and tear on the compressor, since it will need to start less often.

Air Tool	Average cfm
Grease gun	4
1" impact wrench	10
½" impact wrench	4-5
Low-pressure spray gun	3-6
7" angle disc grinder	5-8
Framing nailer	2.2
Orbital sander	6-9
⅜" ratchet	4.5-5
Shears	8-16
Chisel/hammer	3-11

Photograph: Doug Hetherington

An Office Fit for a Shop

The addition of a fully functioning office to their shop not only creates a professional headquarters for the farm business, but also provides a family meeting spot for (from left) Veronica (Andy's wife), Andy, Kevin, Debbie, and Bob Klemp.

A well-designed addition creates the office they always needed

By Dave Mowitz, Machinery Director

When pigs fly. That was when Debbie and Bob Klemp figured they'd add a farm office back when they started farming 20 years ago.

But when their sons, Andy and Kevin, returned to the farm and their operation expanded geographically with some land now 30 to 40 miles away, the Klemps realized they needed a place where the farm team could get organized and plan their hectic days.

"A shop office is one of those things you assume you can always get by without," says Bob. "So you put it off to accommodate other investments such as hog buildings or grain bins, in our case."

But the day came when the Wheatland, Indiana, couple knew that an office was in order.

"We have a large hog operation," says Debbie. "So the guys needed a place where they could change clothes and shower, if needed, when going between buildings in order to avoid transmitting disease. For that reason alone, a shop office was justified."

Armed with a great sense for shop design, supported by input from two sons with a knack for engineering, and equipped with a wealth of construction material purchased cheap at auction, the Klemps set out to create an affordable office to meet a myriad of needs. The result of their efforts not only supplied that office, but also earned them first place in the Best Shop Office category of *Successful Farming* magazine's Top Shops® Contest.

The roots for the Klemps' new office go back to a rudimentary countertop and cabinets they installed along one wall of their existing 48×72-foot shop (see floor plan on the next page). That area certainly was useful to hold service and owner's manuals, parts catalogs, repair guides, and small parts.

"Outside of the wash tub along the wall, there really wasn't a place to clean up when we left a hog facility," Bob points out. "And there certainly was no place to hold meetings or to get a bite to eat."

That inspired the Klemps to add a structure to one side of the shop.

"We figured the best place for a future office would be on the north side of the shop since this was closest to our house and hog buildings,"

Debbie notes. "Plus, this location would place the office in the main traffic area of the farmstead."

The north side of their shop was the location for a storage loft with the previously mentioned countertop and cabinets located beneath that balcony. This became the natural location for a building addition. After much debate, the Klemps opted to add a 24×28-foot structure to house the office.

"Then we got serious about design," Debbie says with a laugh. "We sat around the table and debated our needs, what we wanted, and what we could afford."

Among the list of priorities:

- A spacious, fully functioning kitchen to provide the family and employees a comfortable place to prepare and eat meals.
- A desk for Bob to keep records, monitor markets, and plan farm activities.
- A meeting area for the crew to gather and figure out their daily activities.
- A bedroom where a son or employee could stay the night. That bedroom can also be storage space when not in use.
- A full bathroom with a shower and urinal as well as sink and stool.
- A locker room with laundry for changing/washing clothes worn in the hog facilities.

The Klemps' original shop office is located to the left. The new addition is divided up with an open area that accommodates an office, meeting area, and fully equipped kitchen. Partition walls create rooms for a bathroom, locker room, and bunk area, which can also serve as storage or a future office. A 6×8-foot shed on the back of the office houses an air compressor for the shop as well as the office's water heater and another water heater that supplies the in-floor heating system in the office and shop.

Bob Klemp prefers having his desk in the same open space that houses their office's meeting area and kitchen. Making this space-saving move also "keeps me in touch with what is going on in the rest of the office as well as in the shop (located behind him) and the farmyard," Bob explains. (The farmyard is visible through the window to his right.)

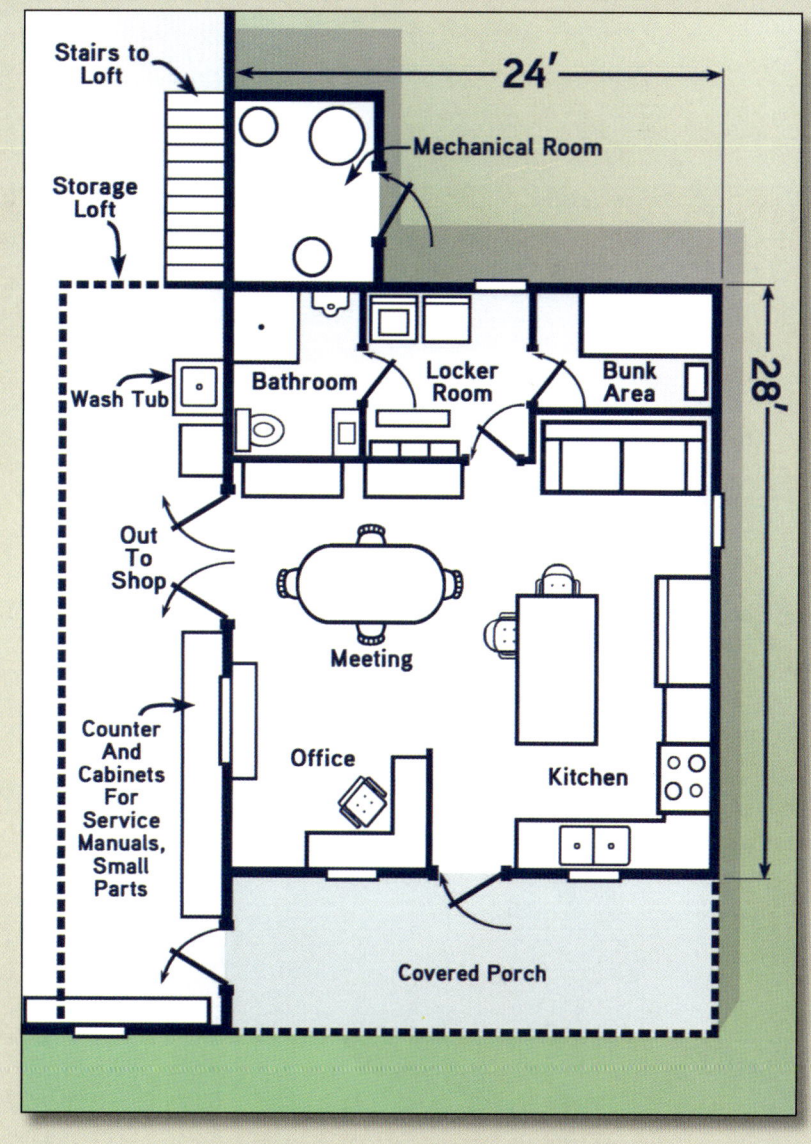

All that and more was squeezed into their final design. Their facility is amazingly spacious considering the amount of room they had to work with. Almost immediately after it was finished, the Klemps noticed how the office became the farm's headquarters.

"Seed and chemical salespeople come in (to the office) rather than our home," Debbie points out. "Even our banker comes out to the office for his visits rather than us having to go see him. This certainly created a professional facility we really needed."

Wow! The banker comes to visit them. It would appear that pigs really can fly! ■

The Klemps built their new kitchen around an island salvaged from a house. The island acts as a table for meals. A fully equipped kitchen surrounds that island. The meeting room is in the foreground. Access to the bedroom (storage), locker room, and bathroom is to the left of the island.

The Klemps' original shop office was located underneath a storage loft located on the north side of the shop wall. The double doors were added to provide access to the new office. The existing countertop and cabinets were kept in place for manuals and parts catalogs as well as to hold small parts.

Shop Office is Essential

Supersized structure acts as this farm's one-stop shop and storage center

By Dave Mowitz, Machinery Editor

When the time came that the Weiss brothers felt they could justify a new shop, the Bay City, Michigan, farm team spent time planning the new structure.

"This was a lifetime investment. So we felt it was crucial to visit other farm shops to get ideas," says Tom Weiss, who farms with his brothers, Dan and Terry.

One conclusion they came to quickly was that their shop needed an office. As shops have become many farms' headquarters, providing office space in that structure makes sense. "When salespeople or other farmers come to the farm, they just naturally go to the shop to find us," Tom says.

When the Weisses developed a floor plan for their 60×60-foot structure, they opted to incorporate space for the office under a 10-foot-deep storage loft that runs across the entire north side of the shop. Furthermore, they positioned the office in the northeast corner of the shop, which is the corner closest to the farmstead's main entrance.

The brothers also opted to keep the office separate from a kitchen. "The kitchen and nearby bathroom were necessities because none of us lives on the farmstead," Tom explains. "But we avoided a kitchen office combination to cut down on traffic. This keeps the office strictly for business and provides a quiet place where we can talk, do bookkeeping, or make marketing decisions." ∎

top: A surrounding laminant tabletop creates space for multiple work areas in the Weiss brothers' office. The window to the left of Tom Weiss overlooks the workbench area. The window behind Tom looks out over the shop's work bay.

bottom: The Weiss brothers positioned their shop's main workbench between their office (located behind the wall to the right of Tom) and the shop's kitchen (located to the left of Tom). This helps restrict lunchtime traffic to the kitchen, keeping the office free of food. It also cuts down on noise.

Farm Headquarters

Jim Cormany intentionally placed his office and break room in a separate area that adjoins his shop. This gives him a spacious office that doesn't steal working area away from the shop.

Shop and office complex serves as the headquarters for this operation

By Dave Mowitz, Machinery Editor

Jim Cormany's first office was typical of any farm – it was way too small.

Offices are like that. Similar to shops, offices seem to shrink in size over time, leaving farmers to ponder why they didn't build larger areas in the first place.

"That certainly was on my mind when designing my shop and office," the Columbia City, Indiana, farmer recalls. "I came from a 10×14-foot office with no windows nor air conditioning. So I wanted plenty of room and better comforts."

Cormany had considered placing the office inside his new shop.

"But I was concerned that this would consume work space on the shop floor," he remembers. "Also, what would happen if I wanted the office to grow in the future? That's a bit difficult to do inside a shop."

Then, too, he needed a break room and work clothes storage for his part-time employees. His solution was to add the office and break room onto the side of the shop. This keeps the rooms separate but adjoining the shop.

"The shop is the hub of many farming operations," Cormany notes. "So it only made sense that the office/break room and shop be kept together."

Cormany's approach to creating a shop-office complex earned him first-place honors in the Best Shop Office category of the Top Shops® Contest. Strategic use of space as well as a provision for future growth allowed Cormany's entry to rise to the top of the competition.

THE MOST COMMON APPROACH

The advantage in having an office share space within a shop is primarily economic. It means the construction costs are lower, and it creates a storage loft above the office. But shop noises, smells, and dirt seem to readily penetrate interior offices. And then there is that matter of future growth.

"True, taking my approach cost more," Cormany points out. "And at first, I worried whether this was

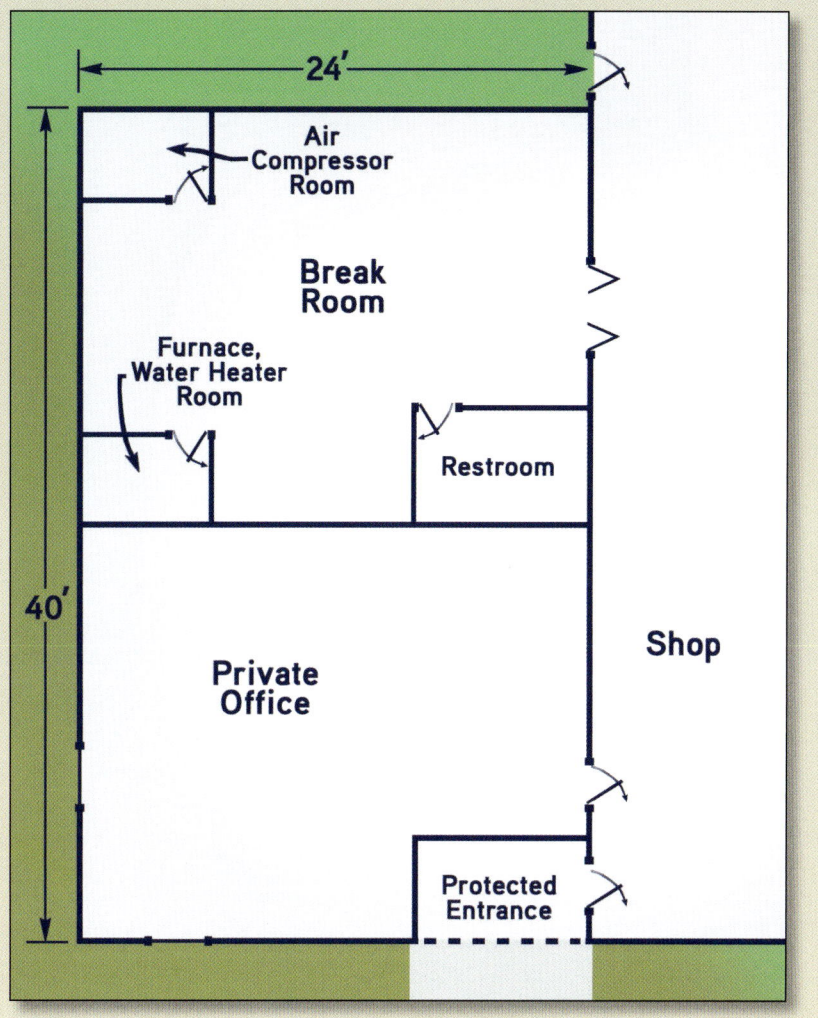

justified. But not any longer. This was one of the smartest decisions I made."

Since its construction, Cormany has discovered that he uses the office much more than in the past.

"It is essential to the operation. All my marketing, purchasing, and business transactions occur in the office," he says. "The break room provides a central location for my part-time help and me to gather and plan out the day."

The cost of the additional structure was further justified by the fact that the break room houses the shop's air compressor, which cuts down on shop noise. It's also where they keep service manuals and shop maintenance records. ∎

Jim Cormany's office is separate from the break room to keep the office clean and to give a professional look. The break room, with a separate entrance to the shop, does double duty housing the shop maintenance records as well as all the service manuals.

"This is where I start every morning and end every day," Jim Cormany says about his office. Cormany's wife assists with field record chores and uses a separate desk located to the right of her husband's main work center.

Cheap Heat

Waste oil burners, such as the Omni model pictured, are a great way to warm your shop. They burn your used oils, cut down on pollution, and save you money over time.

Try an alternative method of heating your space this winter

By Justin Davey

Heating your shop doesn't have to set your checking account on fire. Heating methods like waste oil burners and mixed-fuel burners may be economical and efficient ways to warm your space with resources that are found close to home. Find a model approved by the Environmental Protection Agency (EPA) and enjoy cheap, warm productivity.

WASTE OIL BURNERS

A waste oil burner is a great way to reduce escalating energy prices while efficiently consuming your used oils. They generally pay for themselves in 12 to 18 months. It is figured that 500 to 1,000 gallons of waste oil are all you'll need to save thousands of dollars each year.

Acceptable fuels include vegetable oils, cooking oils, used motor oils, hydraulic oils, used transmission fluid, and combustible synthetic oils, among others. Using these oils also cuts the cost of hauling them to an approved disposal site.

Burning waste oil to obtain cheap heat is also an environmentally friendly way to assist in reducing pollution. By turning a waste product into a valuable fuel, you help conserve natural energy resources and protect water and soil resources from oil pollution. Emissions are similar to burning standard fuel oil, and on-site recycling reduces the chance of accidental spills, improper disposal, and vehicle emissions generated during transport.

Additionally, the liability for proper disposal of used oil extends from the moment it drains into your oil pan until it's processed or burned. This includes any accidental spills or intentional dumping. By burning your used oil in burners meeting EPA standards, you reduce the cradle-to-grave liability.

Similarly, a waste oil boiler heats the used oil under pressure in a closed vessel with the same cost and conservation benefits as a burner.

MIXED-FUEL BURNERS

You may consider using a mixed-fuel burner, which burns wood pellets and corn. Though corn prices are rising alongside propane, natural gas, and electricity, mixed-fuel burners heat your shop with a source that may be nearby.

Clean-burning corn stoves surpass EPA emissions standards and don't create dangerous kerosene buildup. Instead of stalks or cobs, they burn kernels, which contain oil and starch.

Unlike wood-burning fireplaces, corn burners don't need a chimney, though many do install as fireplace inserts. Most exhaust directly through any outside wall through a dryer-like vent.

Once lit, a corn stove can burn for 24 hours without stoking. Corn stoves generally self-feed by an electric auger, and you can set the auger speed to control the heat output. After burning, there will be some residue to remove from the firebox.

To avoid mold, clumping, and smoking, shelled corn for burning should be clean with a moisture content no higher than 15.5%, the same as in your grain bins.

Several manufacturers in the U.S. and Canada make quality, efficient multifuel burners, ranging from simple models to those with a number of high-tech features. Determine your needs based on your shop size and access to fuel.

INFRARED HEATERS

A less economical, yet very viable, method of heating your shop is with an infrared heater. Infrared heaters work the same way as the sun. They are mounted overhead and generate radiant energy, which is converted to heat when absorbed by the objects in its path. The energy is directed toward the floor by highly polished aluminum reflectors.

In a conventional forced-air heating system, the warm air escapes when a door opens. In an infrared system, however, heat remains stored in the objects to help reheat the air quickly once the door closes.

Infrared heaters are either high or low intensity. They use a gas and air mixture to heat either a steel tube (low intensity) or a ceramic surface (high intensity). High-intensity heaters require a higher mounting because of their hotter surface temperature and are used for total building or spot heating. Low-intensity heaters have a lower surface temperature.

When choosing a heater for your shop (most shops are better suited for low-intensity heaters), be sure to consider the burner design, tubing, and reflectors. A swirl burner will help get a better gas and air mixture to push the flame farther down the tube. Choose highly emissive tubing and highly polished, mirror-finish reflectors to push rays downward. ∎

Mixed-fuel burners, such as this model by Baxi, use wood pellets and corn to heat your space. Generally burning for 24 hours or more without stirring, mixed-fuel burners are perfect for heating your shop with sources that may be right in your backyard.

Similar to the sun, gas-powered low-intensity infrared heaters generate radiant energy, which is converted to heat when absorbed by the objects in its path. A gas and air mixture heat a steel tube, and as the energy is emitted, highly polished reflectors push the rays downward.

Waste Oil Heat For Shops

Heidi Otto adjusts the flow of heated water warmed by a waste oil boiler. The Ottos purchased a 240,000-Btu furnace and adapted it to heat water circulated through their shop's floor using a salvaged water coil. "An oil boiler was $12,000, so we saved $8,000 with some engineering," Greg Otto says.

These furnaces offer an alternative, but compare the details before buying

By Dave Mowitz, Machinery Editor

The popularity of waste oil furnaces is evident in the fact that in some parts of the country used engine oil now sells for 10¢ to 25¢ a gallon or more. "It's not always free anymore," says Mark Nicolas of Clean Burn, a waste oil furnace manufacturer. "Yet if you create 700 gallons or more waste oil a year on your own farm, that's enough fuel to heat a large shop."

The major drawback to waste oil furnaces, however, is the high initial cost. Prices for a 100,000-Btu, forced-air furnace without a storage tank start at $3,000, skyrocketing beyond $10,000 for larger (up to 1,000,000-Btu furnaces) and better equipped systems.

Ron and Greg Otto invested around $4,000 in a waste oil furnace, which they adapted to heat water circulated in the shop's floor. "I had hoped it would have paid for itself in two years by not having to buy LP gas," says Greg. "It will take all of four to five years to recoup that cost, however."

The Ottos estimate that they burn 900 to 1,000 gallons of oil a year. Heating their 50×102-foot shop, they use everything generated on their Lester Prairie, Minnesota, farm. "Often we get additional oil from our neighbors," Greg adds.

Waste oil furnaces can also burn #1 or #2 fuel oil. Some systems can handle higher viscosity oils (like gear lube) or fluids with high flash points (like synthetic and vegetable oils). Some furnaces can only burn vegetable oil that has been processed or purified. So this is just one way the vast numbers of furnaces and boilers on the market differ.

COMPARISON POINTS

Some other points to consider when comparing oil burners:

■ **Oil viscosities.** Some furnaces are designed to handle everything from 10W to 50W oils. Other units can't consume thick lubricants.

■ **Single- vs. multiple-pass systems.** Some waste oil furnaces use a single-pass heat exchanger, which is not as efficient as multiple-pass systems that boast efficiencies up to 85% and beyond. That efficiency is often reflected in exhaust temperature. Some furnaces have stack temperatures reaching over 900°F. while others exhaust at 500°F. The difference here is more heat being wasted out the exhaust.

■ **Cleaning requirements.** Some furnaces need only be cleaned once a season. Other products require a cleaning every 50 hours of use.

■ **Warranties.** Coverage and availability of repair parts vary widely across the industry.

■ **Compressed air.** All waste oil burners require compressed air to atomize fuel. Some units come with a built-in air compressor, which likely won't be needed in most farm shops already plumbed for air.

Before buying a furnace, calculate the heating needs of your shop. Any system should provide about 50 Btu's per hour per square foot of shop space, estimates Vern Hofman of North Dakota State University. ■

COMPARISON OF FURNACE FUELS

Fuel	Btu's / Unit of Fuel	Units Needed for 140,000 Btu's	Price / Unit
Waste oil	140,000 Btu's/gal.	1 gal.	5¢/gal.
#2 fuel oil	140,000 Btu's/gal.	1 gal.	$1.80/gal.
Natural gas	100,000 Btu's/ccf	1.4 ccf	$1.80/ccf
Propane	92,000 Btu's/gal.	1.52 gal.	$2.50/gal.
Electricity	3,413 Btu's/kwh	41.02 kwh	6½¢/kwh
Corn	8,500 Btu's/bu.	16.47 bu.	$1.70/bu.

bu. = bushel   ccf = one hundred cubic feet   kwh = kilowatt hour

Photograph: Ron Van Zee

Infrared Space Heaters

Keep warm this winter while you work in the shop

By Laurie Potter, New Products Editor

Photograph: Dave Mowitz

What we complain about during the summer months, we long for once again as the days become shorter and the nights become colder: heat. As the cold weather settles in this winter, you may keep warmer while you work in the shop if you install an infrared space heater.

Infrared heaters work the same way the sun does. They are mounted overhead to direct infrared rays to heat objects such as the floor, people, machinery, and tools, which, in turn, heat the air. In a conventional forced air system, the stored heat (warm air) escapes when the doors are open. But with an infrared heating system, stored heat remains in the objects to help reheat the air quickly once the doors close.

HIGH- AND LOW-INTENSITY HEATERS

Infrared heaters are either high or low intensity. They use a gas and air mixture for combustion to either heat a steel tube (low intensity) or a ceramic surface (high intensity). Both of these heaters can be used with natural gas or liquid propane.

High-intensity heaters require high mounting heights because of the hotter surface temperature. They're used for total building or spot heating. These heaters vent their exhaust products into the space and usually require an exhaust fan.

Low-intensity heaters have an internal fan that fires a flame 2 to 10 feet down a combustion chamber. As the tubing is heated, it emits infrared energy, which is then directed toward the floor by highly polished aluminum reflectors. These heaters have a lower surface temperature.

The majority of today's shops are better suited for low-intensity heaters.

There are three main features to consider when choosing the right low-intensity heater for your shop.

Like the sun, gas-fired low-intensity infrared heating systems generate radiant energy, which is converted into heat when absorbed by objects in its path.

■ **Burner design.** Look for a swirl burner. This design will help facilitate a better gas and air mixture to push the flame farther down the tube.

■ **Tubing.** Some models have a highly emissive black coating on the combustion and radiant tubes, which emits rays away from the tube for optimal output.

■ **Reflectors.** Get best performance out of a heater with highly polished, mirror-finish reflectors, which reflect rays downward.

Before purchasing any system, make sure the clearances to combustibles are taken into consideration. This is the minimum distance that must be maintained between the tube heater surface and any combustible product.

Depending on the application and model, it's recommended that mounting heights be 10 feet or higher.

TWO-STAGE HEATERS SAVE SOME FUEL

Two-stage heaters are becoming more popular as fuel costs go up. They utilize a two-stage gas valve that allows units to operate in high- and low-fire modes. This reduces the on/off cycles and can give up to 12% fuel savings over a single-stage heater.

Most low-intensity heaters are priced by tube length. Base cost of a standard, single-stage, 40-foot tube heater is about $1,000, depending on the features. ■

Battery Chargers

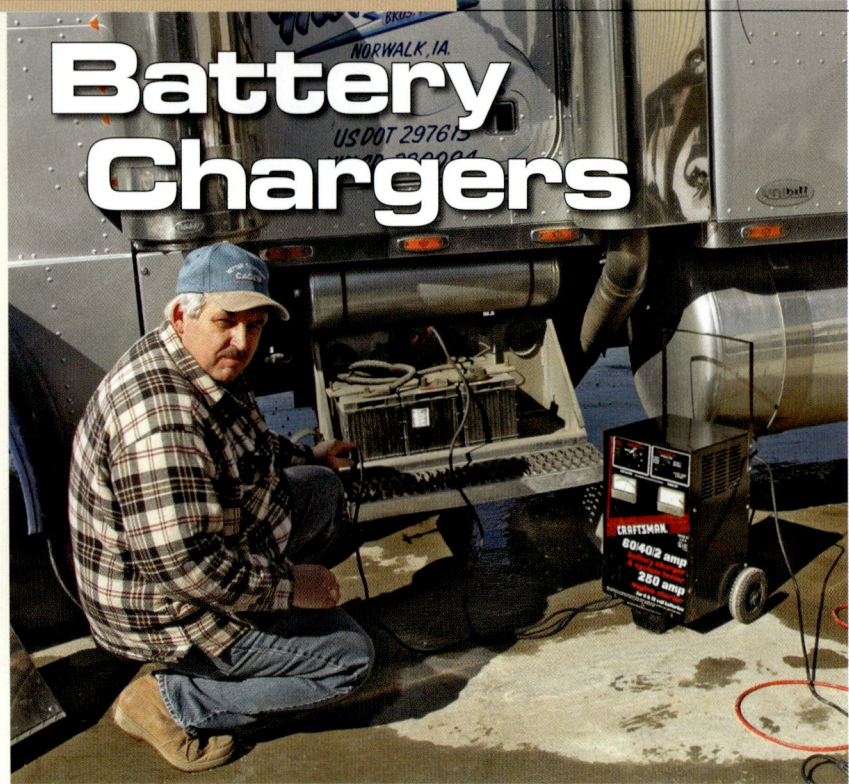

This battery charger sees regular duty on the Dick (left) and Bob McPherson farm near Norwalk, Iowa. The brothers have 23 vehicles on the road or in the field. (Note: Some battery chargers are designed for outdoor use and some are not. As a precaution in damp situations, use a ground fault interrupter or plug the charger in inside. Furthermore, protect the charger from moisture getting into the case.)

These chargers are big enough to start tractors and trucks

By Rich Fee, Crops and Soils Editor

One of the most aggravating sounds on a farm is an engine that won't quite turn over because of a weak battery.

But having the right kind of battery charger in the shop can reduce the level of aggravation and the amount of valuable time lost.

When it comes to battery chargers, size matters. A 10- to 50-amp charger is fine for automobiles and yard equipment. But when you start talking tractors and trucks, it's nice to have a charger with 200 to 300 amps of cranking power. That's big enough to get a tractor going in a hurry (unless a battery is shot). Plus, it can charge batteries faster.

SIZE IT TO YOUR EQUIPMENT

Pete Maziarz, customer service manager for Schumacher, a major manufacturer of battery chargers, says, "The cranking amperage of the starter motor in your machine is the determining factor on how big a charger you need and what kind of draw load a battery is going to see."

Wheel chargers have about 250 amps of cranking power. But when shopping, don't automatically rule out some of the 200-amp models with a handle instead of wheels. They can be really convenient as long as 200 amps is enough cranking power. Plus, they are less expensive.

As you would expect, there are some significant price differences among brands of battery chargers. Perhaps more surprising is the wide range of prices among retailers for the same brand and model.

Online retailers often have lower prices because they have very little overhead. But they aren't equipped to provide service like a storefront dealer is. Only you can decide how important that is for your operation.

BY THE NUMBERS

Most chargers handle 6- and 12-volt batteries. Some also handle 24-volt batteries.

Often there are several numbers printed on the case. If the largest number is 250, that typically means it has 250 amps of cranking power for 12-volt batteries.

The lowest number is usually 10. That means it has a 10-amp setting for slow charges, which big batteries sometimes need. However, some big chargers also have a 2-amp setting for charging motorcycle and lawn mower batteries.

Features account for some price differences. Most chargers have ammeters, which show you the amperage being forced into the battery.

Voltmeters are less common. They indicate the state of charge of a battery. Randy Judge is the sales manager for the NAPA distribution center in Des Moines, Iowa. He believes that opting for a voltmeter "is money well spent."

Most battery chargers come with fairly detailed operating instructions. There is also a wealth of information available online. For example, a John Deere site (http://jdparts.deere.com/partsmkt/document/english/pmac/5666_gn_JDLead_AcidBatteryChargeProcedure.htm), has a useful six-page pamphlet on charging procedures for batteries. ∎

Photograph: Rich Fee

Battery-Powered Grease Guns

A.J. Blair, Dayton, Iowa, likes the speed and easy handling of a cordless battery-powered grease gun on machinery with many zerks, such as a tillage tool.

Speed, simplicity, and ease of operation are perks when greasing zerks

By Gil Gullickson, Crops Technology Editor

Dirty duds and aching muscles from morphing into a human pretzel to reach zerks used to be part of greasing machinery. Fortunately, today's cordless battery-charged grease guns nix the grease spills that accompanied old manual guns. Ditto for user-friendly design and flexible hoses that key easier greasing in tight spaces. These units also speed greasing in today's tight time windows.

FINDING A FIT FOR YOUR FARM

"For something that has a lot of grease zerks like a tillage tool, cordless grease guns are really nice," says A.J. Blair, a Dayton, Iowa, farmer.

"After 20 pumps, your hand gets tired. If you only need to give a few shots to machinery that doesn't have a lot of zerks like a combine, hand pumps are almost easier to use."

If you do enough greasing to justify purchase, there's a good range of options on the market.

"The performance factors that are important in the marketplace are pressure, volume, battery life, and gun performance," says Americo dos Santos, Alemite product manager.

Adequate pressure is paramount for breaking through clogged zerks.

"Particularly in agriculture and construction, grease fittings collect so much dirt and debris," says Keith Rohan, product manager for Lincoln Industrial. "With a manual grease gun, it requires a lot of muscle to break through a blockage or to grease a tight bushing. You can physically wear yourself out by the time you grease 30 to 40 zerks."

Not so with battery-powered grease guns. Maximum pressure that ranges between 7,500 and 10,000 pounds per square inch (psi) can break through sticky clods.

These guns also deliver up to five

times the volume of manual models, which speeds the greasing process.

Battery life and long-lasting recharges also eliminate downtime. Twelve- to 18-volt NiCad batteries (or two) and a battery charger accompany cordless battery-powered grease guns. Industry officials say the time between recharges varies between three and 10 14-ounce grease cartridges. Battery life and time between recharges hinge upon several factors.

"Under high-pressure applications, the battery won't last as long as under low-pressure applications," says Dave Babics, senior product manager for Plews & Edelmann. "Temperature comes into play, too. Grease thickens and is harder to pump when temperatures are in the high 30s and lower 40s."

BUYING CONSIDERATIONS

Lower-priced versions contain a more basic package, with a 12-volt battery or two, a one-hour charger, flexible hose, and up to 8,000 maximum psi.

Higher-priced models offer features like two-way speed switches.

"If you put it on a high-pressure setting, you can break through those tough-to-grease fittings with 7,500 psi," says Rohan. "With clean lube points, you can switch to the high-volume side and speed through the lubrication process."

Other grease guns feature variable-speed triggers. "With variable speed, you can go 10% of full speed all the way up to full speed," says Richard Sharpless, managing director for Lumax Lubrication.

Many models can use grease from cartridges, bulk load, or suction load. They also have safeguards to prevent grease blowups and spills and to keep grease flowing into zerks.

Locking rods prevent grease spills by securely locking the grease cartridge in the gun. Trigger locks

"Lubrication protects the machinery and equipment investment of farmers," says Keith Rohan, product manager for Lincoln Industrial. "But it takes away from time in the field. The primary benefits of these tools are speed and efficiency so farmers can shorten downtime."

Remember when reloading a grease gun could result in a grease eruption? Meanwhile, air pockets may have prompted you to bang the gun on the frame of the implement you were greasing. No more. Today's battery-powered grease guns have features that prevent these malfunctions.

can prevent accidental grease release, while internal pressure release valves allow excess grease to flow back into the barrel.

Cordless battery-powered grease guns feature user-friendly grips and 2.5- to 3.5-foot-long flexible hoses for greasing hard-to-access zerks. Models contain ways to exit air from grease, such as air bleed valves. ■

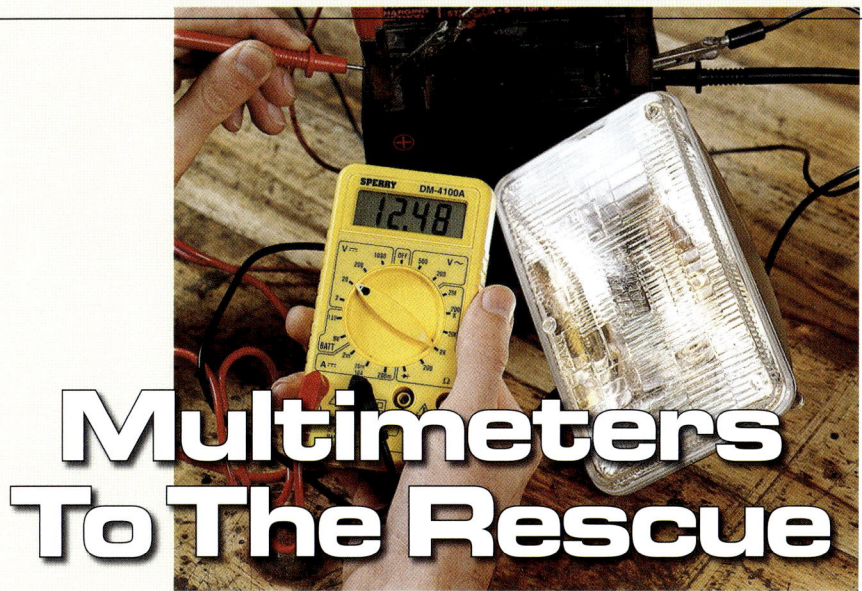

Multimeters To The Rescue

This electrical tester can diagnose many common engine problems

By Steven Parks

Photograph: Ron Van Zee

A multifunction test meter is a powerful weapon in the mechanic's battle with electrical problems. But it's often tucked away and forgotten because the owner never learns to use it. An intimidating tool at first glance, the multimeter has so many different functions available and only brief, cryptic labels. But for most mechanical work, you'll need just two measurement ranges: DC voltage from 10 to 15 volts, and resistance from 0 to 20,000 ohms.

The right meter for the farm is a simple, rugged meter built for the automotive mechanic trade. Desirable features include auto-ranging ability, slip-on alligator clips for the test probes, audible continuity checker, and automatic power-down when not in use. Such meters range in price from $29.99 to $45 and are available at most automotive tool and parts outlets.

CURING A DEAD BATTERY

Select DC Volts on your tester, with a maximum scale reading that accurately measures from 10 to 15 volts. Choose a maximum scale reading of 15 to 30 volts, depending on what is available on the meter. If it's an auto-ranging digital meter, don't worry about choosing the measurement range, just set it to measure DC Volts. A digital meter gives precise measurement.

First, check the battery voltage with the engine and accessories switched off. Touch the red test lead firmly to the positive post on the battery and the black lead to the negative post. Get a good connection because accuracy depends on it. You should have 12 to 13 volts. Less than 11.5 volts and you may have to charge that battery or jump-start the vehicle to get the engine running for the following test.

Next, with the engine running at fast idle and all accessories switched off, check the battery voltage again. It will be 13.5 to 14.5 volts if the alternator is charging the battery. If it is the same or lower than your engine-off reading, the battery is not being charged. If it goes higher than 15 volts and stays there, your voltage regulator is defective and should be replaced.

Resistance is just what it sounds like: the amount of restriction in the

Just two settings on a multimeter – DC voltage from 10 to 15 volts and 0 to 20,000 ohms – will diagnose many basic engine electrical problems such as batteries not charging or engines running rough.

flow of electricity through a circuit. It is measured in ohms. Analog-type meters, where the reading is displayed by a needle swinging across a scale, are the easiest to understand when measuring and comparing resistance.

All modern ignition systems use resistor-type spark plug wires to reduce Radio Frequency Interference (RFI). These wires may develop too much resistance as they age, leading to hard starting, rough running, and missing under load. This problem is easy to pinpoint with a multimeter.

Set your meter to a maximum reading of around 20,000 (20K) ohms. With an auto-ranging digital meter, just set it to measure ohms – you don't have to manually select a range. On most analog-type meters, you'll want R×100 or 100 times the indicated reading on the meter face. Results may vary, so consult your meter's documentation. Calibrate an analog ohmmeter by touching the leads together and adjusting the calibration thumb-wheel until the meter reads zero. Digital meters generally do not need calibration before use.

RESISTANCE IN PLUG WIRES

Remove one plug wire at a time, so there is less chance of reconnecting incorrectly. Get a good connection between your test leads and the ends of the wire being evaluated. Don't forget to check the coil wire, too.

Normal plug wire resistance runs from 3,000 to 7,000 ohms per foot of length, so it's normal for longer wires to have more resistance than their shorter neighbors. Replace any wire that has more than 15,000 ohms resistance, unless the manual specifies otherwise. If several wires are bad, it's wise to replace the whole set. ∎

Tire Pressure Gauges

Proper inflation rates key to optimum tire performance

By Laurie Potter, New Products Editor

Tractor tires are under a lot of pressure. Not only do they have to handle the load of the tractor, but also they probably need to support an implement. With this much responsibility riding on their rubber shoulders, it seems natural that tire owners would check inflation rates on a regular basis.

After all, an improperly inflated tire runs the risk of compromising the life of the tire.

"You wouldn't run your tractor engine a couple of gallons low on oil, so why would you run your tires 2 or 3 psi lower than recommended?" questions Ken Brodbeck of Firestone.

He says if operators properly inflate tires, they will:

- Lengthen the life of their tires.
- Experience excellent traction.
- Lower wheel slip.
- Spend less time in the field.
- Use less fuel.
- Experience a softer ride.

"Fuel economy is very much affected by tire pressure. So in this era of very high fuel prices, it's important to have tires properly inflated," says Brodbeck.

SELECTING A GAUGE

To gauge tire pressure, you need an instrument that provides an accurate reading. There are three main types of gauges: pencil, dial, and digital.

"I've used a pencil and digital gauge, and they've both been accurate,"

With the press of a button, digital tire pressure gauges record the psi of a tire to check for appropriate inflation rates.

A properly inflated tire is designed to operate with flexibility in the sidewall. Too little or too much air can have a negative impact on traction, flotation, productivity, tire wear, and tire life.

Overinflated tires are a common cause of poor tractive efficiency and compaction. Sidewalls on grossly overinflated tires are stiff and more likely to experience small breaks if they hit an object or rut.

Brodbeck says. "The nice thing about the digital gauge is that it will hold and record pressure value until you can take a look at it. Some pencil gauges can't do this."

He says the key thing about gauges is to choose one with the appropriate psi rating for your tires.

"Don't use a 100-psi truck gauge when checking tractor tires. Choose a gauge that has a range of 0 to 30 psi for large tractor tires. If checking sprayer tires or other high-pressure tires, use a 0- to 80-psi gauge," Brodbeck advises.

Pencil gauges

A pencil gauge is made of a plastic or metal casing, has a pocket clip, and a deep-set chuck, usually with a pressure-release bump on the opposite side. The amount of pressure is shown on an indicator bar that extends when pressure from the tire is applied to the gauge.

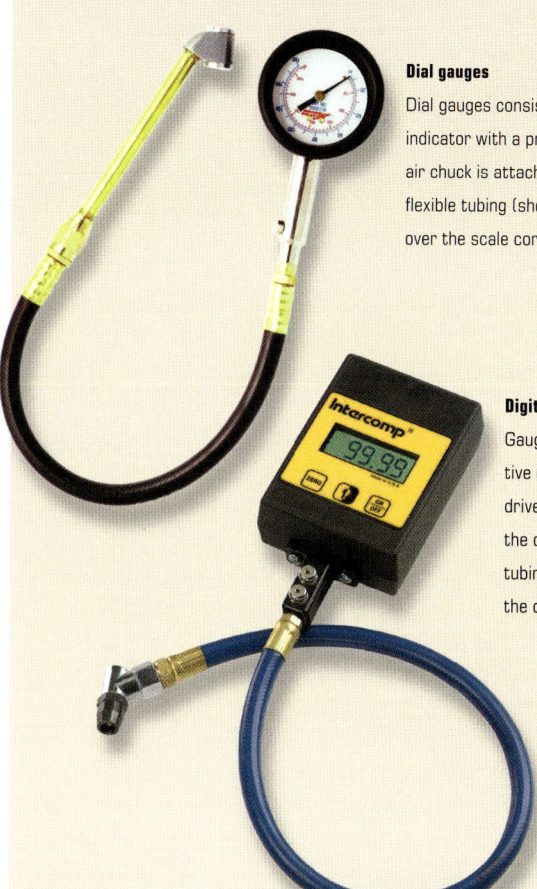

Dial gauges

Dial gauges consist of a protective casing around a dial indicator with a pressure scale on the dial background. An air chuck is attached to the dial rigidly or with a section of flexible tubing (shown left). The indicator needle rotates over the scale corresponding to the pressure in the tire.

Digital gauges

Gauges with a digital readout have a protective casing with the electronics necessary to drive the readout. An air chuck is attached to the casing rigidly or with a section of flexible tubing (shown above). Turn the gauge on and the display indicates pressure.

PROPERLY CARE FOR YOUR TIRE GAUGE

As tire gauges are used, they are exposed to moisture and foreign materials. Protect your gauge by keeping it out of the toolbox and away from the elements.

"Two of a gauge's biggest enemies are getting calcium chloride [tire fluid] in it or getting dropped," Brodbeck says.

He suggests not using the same tire gauge for your pickup tires as you do for your tractor tires. Use one gauge for air-filled tires; use another gauge for calcium chloride-filled tires.

Brodbeck recommends having at least two gauges so you can check them against one another. "And don't be afraid to throw a bad gauge away," he says.

For under $20, you should be able to buy a decent gauge, Brodbeck says. "You don't have to buy a $50 to $100 gauge to get a reliable, accurate reading."

KNOW YOUR PSI

Farmers can err in both directions on tire pressure – way too high or way too low. "Either one can be costly," Brodbeck says.

Be sure to always check tire pressure when tires are cold. The rule of thumb is that for every 10°F., a tire's pressure changes 1 psi. Tractor tires will increase 1 to 2 psi when they're being run during the workday.

"That's why it's important to check pressures once a week and to know what the inflation pressure is supposed to be for your tires," Brodbeck explains. ∎

5 STEPS TO OPTIMAL PRESSURE

1 Determine the front and rear axle weight and the weight per tire.

2 Figure total tractor weight and the number of pounds required per hp. to efficiently transmit wheel torque.

3 Check to make sure the tire size is right for your engine's hp.

4 Calculate the proper weight distribution required for both the front and rear axles. You may find that you need to add cast weights to get the weight properly distributed.

5 Visit your tire manufacturer's Web site to determine the correct inflation rates. ∎

John Dappert, Oblong, Illinois, has compiled a list of his top 10 shop tools.

Top 10 Farm Shop Tools

A practical guide to must-have tools for any farm shop

By John Dappert

Years ago, in a simpler time, our farm had an excellent machinery dealer only 3 miles west of the farm, and our tool shed was very basic. Oil changes and adjustments were about all we did back in the 1950s. Most other repairs were done by an excellent local welder and repair shop.

The local welder left, and the dealer lost his franchise. That's when our farm began to accumulate tools to do more of the repairs ourselves. Following is a top 10 list of basic tools I've accumulated and use often.

1 Big air compressor. I bought several puny compressors that wore out until we finally broke down and bought a good one. I now wonder why we ever tried to save money on that item. The larger compressors were very handy for dusting off radiators and machinery, as well as doing basic painting. Air tools add a lot to my productivity, locking down nuts, for instance, much more tightly than I can by hand.

You can get carried away accumulating tools that go with a good compressor since there are so many available. A good set of impact sockets is necessary to keep from breaking all of your regular sockets. I even went up to a 1-inch drive impact and sockets when I bought a semi truck. As air over hydraulic jacks became more reasonably priced, I added a 12-ton and a 40-ton bottle jack with air controls.

2 Vise. I bought several cheap vises that seemed to be heavy-duty, but they all broke quickly. You need a good Wilton. Buy nothing of lesser quality. All I know for sure is that just because a vise looks tough doesn't mean it will stand up to heavy use. There is a big difference in materials, and this is an example of when price, indeed, determines quality.

A good anvil is handy, too. But heavy pieces of iron can do much of the work of an anvil if a good one is not available.

3 Workbench. If you really want a good workbench, you should make it yourself out of at least 2-inch lumber on the top, and brace it strongly across the sides. My workbench has 4×6-inch legs. It's best to put on a steel plate top, at least where the tabletop vise is bolted on, to hold the vise more securely. Adding a full steel plate on much of the surface was helpful, but I was advised by a professional mechanic in the family to leave at least half of the bench surface wood. Small, dissembled parts seem to be easier to work on with a wood surface. Furthermore, precision parts don't get damaged as easily on wood.

4 Bolt bins and a good selection of bolts. I prefer the bolt bins that are enclosed. I've found that open-topped bins are a great haven for birds. If your doors are tight and you keep them closed, however, the open-top bolt bins are much easier to use. And they are more convenient when it comes to taking inventory.

But no amount of bins is useful if you let those bins get empty. We get by with a set of bolts and nuts in 1", 1½", 2", 2½", 3", 3½", and 4" lengths.

5 Drill press. The drill press was one of the later additions to our shop. A handheld drill can do the job in many cases. But I've found that a drill press is much more precise and easier to use, especially for drilling large holes. The expense for a press can be delayed until more essential tools are purchased, but it is a much-used tool in most shops once installed. Often, you can purchase larger surplus presses from a factory for reasonable prices. If you can find one, such a tool increases the amount of speed control, which is useful in drilling different size holes.

6 Hydraulic press. Hydraulic presses can make pressing bushings and bearings a much easier job. A 30-ton press is great because it can be used to straighten out many larger pieces. There is no place to stop on the press size – the larger the better. My 12-ton press got me by, even though I had to beef up the cheap thing several times. I added an air jack to replace the pump one. This saved a lot of time repositioning materials.

7 8-inch angle grinder. A 4-inch grinder is handy to use when you don't want the bulk of the big one. But 8-inch and larger angle grinders are essential when preparing welds and sharpening mower blades. Bench grinders are also handy. Nothing seems to beat the portability of the handheld side grinder.

8 Welder. The little 225-amp buzzboxes will do for some of us. But if you do more welding, the duty cycle of about 20% on a smaller one soon slows you down. You'll need a bigger model with a higher duty cycle if you're a person who does quite a bit of welding.

Like anything else, the consumable materials (such as rods) should be on hand for emergencies. Storage of these rods is a problem since humidity often can ruin them.

The wire (MIG) welders are much easier to use, and the choice of such a machine is probably much better than an old stick welder.

9 Chop saw. The chop saw was another tool that took me a long time to add to my shop. The gas torch does work on cut-off jobs, but the chop saw is so much cleaner and faster than the torch. With a metal cut-off saw, you have no slag and better fit, preparation of a cut piece of steel is much easier, and the cut usually makes for a stronger weld.

I regret not having purchased the chop saw sooner.

10 Torch. A gas torch is still a great tool for any shop, especially if equipped with a good rosebud heating tip. There are so many times equipment must be heated during repair. The various size tips can also be handy for bronze repairs and small sheet metal welding.

OTHER TOOLS TO CONSIDER

That's my top 10 list. Of course, the socket sets and end wrenches, bearing pullers, tape measures of all sizes, and all those basic tools are a given in a must-have list of tools.

There are so many little things like drills, woodworking equipment, all kinds of saws, soldering irons, an electrical crimping tool with various tips, a good electrical tester, and heavy extension cords. These are but a few small tools needed besides the basic hammers and screwdrivers and saws.

I was surprised how much I used a forklift in repairs after I got one to handle seed. Besides lifting things like feeder houses on combines or using it as a platform to reach and safely work on high places of the combine and other large equipment, the forklift was a big plus. The smaller forklifts get around a crowded shop very well, and they can take the place of a jib crane for lifting. ◼

Your best bet is to invest in an 8-inch or larger angle grinder.

Building your own workbench provides the opportunity to create a stout-for-the-farm structure.

Every shop should have a stick welder. But MIG machines have also become a must for farm shops.

YOUR TOP TOOL

Participants in an Agriculture Online® poll were asked to rank their single most important shop tool. Here are the results.
- ◼ Air compressor: 63%
- ◼ Welder: 21%
- ◼ Torch: 9%
- ◼ Crane or hoist: 1%
- ◼ Other: 4%

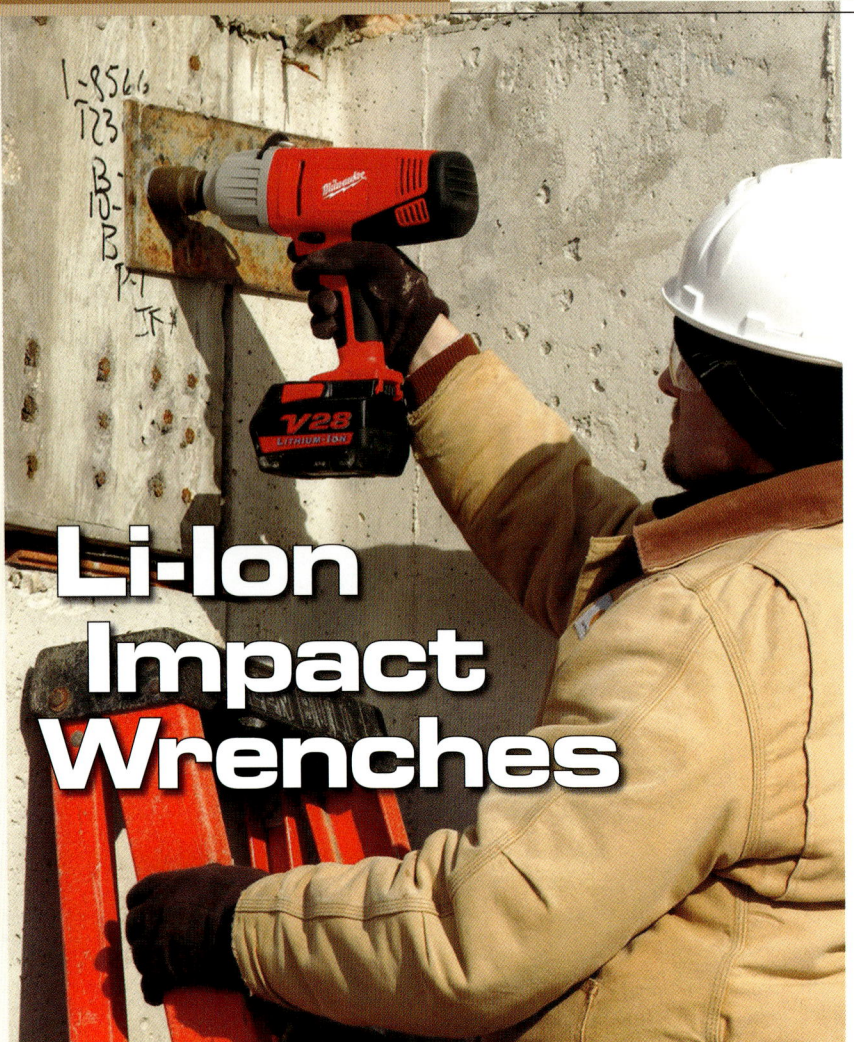

The energy density of Li-Ion batteries allows them to store more energy at less weight. Thus, a Li-Ion tool delivers more torque at a third of the weight of a NiCd or NiMH tool.

Li-Ion Impact Wrenches

Lithium-ion battery technology is poised to take over the cordless market

By Dave Mowitz, Machinery Editor

Few companies can claim they revolutionized an industry. Milwaukee Electric is one of those few. In January 2005, they launched a line of cordless tools powered by lithium-ion batteries – the same technology used to power cellular phones and laptop computers.

Since then nearly every cordless tool company has responded by introducing its own lithium-ion, or Li-Ion, tool line. A quick count on the Internet reveals 62 Li-Ion tools on the market. And predictions are that all cordless tools will run on Li-Ion batteries, as opposed to nickel-cadmium (NiCd) and nickel metal hydride (NiMH), by the end of this decade.

HIGH-ENERGY DENSITY

Li-Ion's advantages over older battery technologies are numerous, which explains its instant popularity. Batteries made from the lithium metal base have a higher energy density and, thus, a greater ability to store electricity per ounce of battery weight. Li-Ion's gravimetric energy density ranges from 100 to 135. In comparison, NiCd density ranges from 45 to 80, and

NiMH's density is 60 to 120.

Then, too, Li-Ion cells operate at higher voltages than other rechargeables. Typically Li-Ion batteries discharge at 3.6 volts vs. 1.2 volts for NiCd or NiMH. This means a Li-Ion tool can operate with a single battery cell as opposed to multiple-celled NiCd or NiMH batteries. This accounts for Li-Ion's weight advantage. For example, Milwaukee's 28-volt Li-Ion battery delivers up to twice the runtime and 40% to 50% more power than an 18-volt NiCd Milwaukee battery of the same weight.

Another Li-Ion advantage is that they maintain their power output throughout their entire discharge cycle. Plus, Li-Ion batteries have a lower self-discharge rate. NiCd or NiMH batteries can lose from 1% to 5% of their charge per day even if the battery is not in the tool. Li-Ion batteries will keep most of their charge months in storage.

MORE EXPENSIVE FOR THE TIME BEING

The major drawback to Li-Ion tools is their cost, which is around 30% to 40% higher than a comparable NiCd tool. This difference is reflected in the fact that Li-Ion batteries are more complex to manufacture. Li-Ion cells require special circuitry that limits peak voltage during charging or prevents voltage from dropping too low on discharge. These technical requirements mean you can't use a NiCd or NiMH charger for a Li-Ion battery. The industry is, however, working on Li-Ion chargers that are "backward-compatible" and thus can charge NiCd or NiMH cells. That will help absorb some of the sticker shock of purchasing a Li-Ion tool.

PRICES EXPECTED TO DROP

Then, too, the cost of Li-Ion tools is expected to drop as more are manufactured. And in that regard, the proliferation of Li-Ion-powered products is amazing. Consider that no Li-Ion impact wrenches were available two years ago. Five models are now on the market. Figure on that number to double by year's end.

Other firms selling Li-Ion tools (but not yet impact wrenches) include Bosch, Hilti, Metabo, Panasonic, Ridgid, and Sears Craftsman.

Impact wrenches are the one cordless tool that farmers have found to be nearly indispensable for their operation. For example, Darrell Geisler got a Milwaukee NiCd cordless impact three years ago. He has since purchased an additional cordless impact wrench so he can have one for his pickup and one in his tractor. "Having one in the field makes adjusting a planter, for example, so much easier and faster," says the Bondurant, Iowa, farmer.

In that regard, the advantage to Li-Ion impact wrenches goes beyond their greater voltage oomph to loosen a rusty 1-inch nut on a mounting bracket. Such a tool can also lie in a tractor or pickup cab for months and still keep a charge.

MUCH LONGER WARRANTIES

Another farmer-friendly advantage the Li-Ion revolution has produced is a longer warranty. As noted in the table below, both Hitachi and Milwaukee offer five-year warranties on their tools as well as their batteries – coverage that was unheard of two years ago.

The one-year warranty is still common on many NiCd and NiMH batteries. ∎

Light It Up

Brad Minor's shop near Rutland, South Dakota, is well illuminated thanks to the white metal sheeting on the ceiling that reflects light from 34 high-pressure sodium light fixtures.

A well-lit shop is a safer and more efficient work area

By Laurie Potter, New Products Editor

Relying on one type of light to be all things to each area of your shop may result in high energy costs. One of the biggest misconceptions about shop lighting is that it uses an insignificant amount of energy, says Craig Metz, vice president, EnSave, Inc.

"A good lighting design can save hundreds or even thousands of dollars over the life of the fixture compared to a conventional lighting design. When farmers make an effort to conserve energy, it adds up to a big reduction in energy use and pollution," Metz says.

Not only can energy costs be reduced but also frustration levels can be minimized, especially when working in a dimly lit corner.

LIGHTING CHOICES

There are a number of lighting choices such as high bay, low bay, T5 and T8 fluorescent. So how do you decide which type is best?

According to Bill Balfour, PFO Lighting, there are three things to consider when lighting a shop: ambient temperature, mounting height, and general vs. task lighting.

1 Ambient temperature. Even though fluorescent lighting is the most energy efficient, it doesn't do well in cold temperatures.

"If the ambient temperature is 40°F., T5 fluorescent lights will lose 75% of the output, and T8 fluorescent lights will lose 40% of their light output," says Balfour.

Metal halide lights are unaffected (to about -20°F.) because they are self-heating.

2 Mounting height. Balfour says a height of 8 to 14 feet can usually accommodate a T8 strip-type light (traditional channel fixture). A T8 high-bay fluorescent-type fixture is recommended for higher mounting (14 to 20 feet). For above 20 feet, he suggests a T5 high-bay fluorescent.

"The high-bay fluorescent fixtures have better reflectors to focus light down from a higher mounting position. In metal halide, I suggest a 250-watt fixture for mounting from 12 to 20 feet high, and 400-watt fixture above 20 feet," he says.

3 General vs. task lighting. How long will the lights be on at a time? For example, metal halides like to stay on and not be turned on and off. Balfour recommends that you make general lighting sufficient for your needs and add task lights for specific locations where needed.

If possible, put banks of lights on separate switches. In doing so, you won't have to turn on all the lights just to grab something from one area of the shop. Also, have a separate breaker for lights in case a tool should trip a breaker. ∎

Lift It Up

Joel Jakobs, Milledgeville, Illinois, uses the electric chain hoist in his shop for routine maintenance of his equipment as well as repair work.

Repairs and maintenance are easier with a shop hoist

By Laurie Potter, New Products Editor

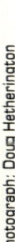

For Joel Jakobs, Milledgeville, Illinois, the hoist in his shop has become an integral part of his operation. "The versatility of it is great. There's not much we don't use it for," he says.

His electric chain hoist from Budgit can lift 3 tons. The frame is a homemade unit with legs that telescope up and down.

"We wanted the frame to be wide enough to clear any piece of equipment we have," says Jakobs.

CHOOSING A HOIST

The amount and type of equipment you have will determine whether or not a hoist is an investment worth making. Hoists range in capacities. As you consider a hoist purchase, take into account capacity, lift, suspension, and the type of trolley that works best.

■ **Capacity.** Determine the maximum load to be lifted. If the load falls between a standard rated capacity, always go with the higher capacity. For example, if the largest piece of equipment you will lift has a weight of 4,000 pounds, buy a hoist with a 3-ton (6,000-pound) capacity.

■ **Lift.** The lift of the hoist is the height needed to raise a piece of machinery. To determine total lift, measure the distance from the bottom of the overhead beam to the lowest point on the floor to be reached.

■ **Suspension.** There are two types of suspensions for a hoist. A hook-type suspension allows the hoist to be hung just about anywhere and is used when a hoist needs to be moved easily to other locations. The lug-type saves headroom and is used with rigid-mount trolleys or when you want to permanently mount a hoist in a fixed location.

■ **Trolley.** There are three standard trolley types that can be used to suspend hoists.

1. A push trolley is best for light capacities and lifts below 20 feet.

2. For higher capacities, a hand-geared trolley is a good choice. It offers the most precise control for load spotting and is recommended where lifts are above 20 feet.

3. A motor-driven trolley is the most widely used method of suspension, especially with capacities over 2 tons.

Jakobs says that for under $1,000 (his cost includes the used hoist and a self-built frame), it's been well worth the investment. ■

Photograph: Doug Hetherington

Floor Jacks

Know how much your floor jack can lift safely before your pickup, tractor, combine, or any other piece of equipment ever leaves the ground.

Before you lift, know your jack's capabilities

By Laurie Potter, Deputy Machinery Editor

Caring for the equipment around your farm is a never-ending job. So before you delve into maintenance chores and elevate another vehicle or tool, know how much your jack is capable of raising off the ground.

The size and type of jack you choose will be determined by the weight of what you're lifting.

There are three basic types of floor jacks: hydraulic, air hydraulic, and mechanical. Hydraulic and air hydraulic jacks are the most popular for obvious reasons – they help you lift a piece of equipment quickly and effortlessly.

"Jacks that are air and hydraulic are made for very, very heavy lifting. These type of jacks are good for combines and tractors. They can actually hook into an air compressor system instead of having to pump and pump and pump," says LeRoy Allen of Northern Tool.

WHAT TO LOOK FOR

According to Allen, there are key considerations to keep in mind when shopping for a floor jack.

"The main thing to consider is what you'll be lifting with your jack," he says. Once you determine what you want to raise, you can narrow your choices to the appropriate weight size. Every floor jack is rated for different weights, and jacks generally range from 2 to 20 tons.

Another consideration is whether or not you'll have access to an air compressor. "A lot of people seem to have to go out to the field for repairs. Then they get there and don't have an air compressor with them," Allen says.

He also suggests purchasing a jack that is wider for more stability. "A wider jack is a lot safer since it can take vibrations better," he notes.

The surface you're lifting off of is also key. "A cement base or hard ground to work on is best," he says.

STEEL OR ALUMINUM

Jacks are made of either steel or aluminum so you'll need to consider which is best for your use. "Even though aluminum is a lot lighter to haul around, I recommend an all-steel unit because it's more durable," Allen says.

Be aware that some models include a Quick Lift design, which brings the jack up to the bottom of whatever you're lifting. That makes the process a little quicker.

THE BEST POSSIBLE WEIGHT FOR THE JOB

"For repairs around the home, a 3-ton unit should be sufficient. But it can depend on the type of equipment and vehicles you own," Allen says. "For farm use, that's where you get your long-arm jacks and get into 5- and 10-ton units."

Five- and 10-ton jacks are heavy. A 5-ton jack weighs about 246 pounds and a 10-ton jack weighs about 350 pounds. Allen recommends keeping these in your shop.

No matter what you're lifting, be sure to have jack stands and blocks on hand.

MAINTAINING YOUR JACK

As with any piece of equipment around the farm, you want to maintain it properly for longer life. A floor jack is no exception. To get the best performance and longest life from your jack, replace all fluids at least once a year.

For hydraulic jacks, use a good grade hydraulic jack oil. Avoid mixing different types of fluid. Never use brake fluid, transmission fluid, turbine oil, motor oil, or glycerin. If you use an improper fluid, it can cause the jack to fail, leading to the unwelcome potential for a sudden loss of load.

Pivot points, axles, and hinges should be lightly lubricated periodically to prevent rust and assure that wheels, casters, and pump assemblies move freely.

When not in use, store the jack with the saddle fully lowered. ■

Photograph: Harlen Persinger

Make It A Clean Sweep

Photographs: Ron Van Zee

Columbia City, Indiana, farmer James Cormany tests out vacuum suction at his shop.

The working end of a shop vacuum consists of curved vanes sandwiched between discs. This creates impellers that act alone (single stage) or in combination with two or three impellers (double or triple stage).

Performance differences separate shop vacs

By Ron Van Zee

While all the gadgets that come with utility vacuums are impressive, they don't reflect the true measure of such a machine's ability to suck debris off a shop floor.

And surprisingly, the size of a vacuum's engine isn't a good guide of its strength. Electric motors on shop vacs are sometimes rated above their true work strength. This is often the case with peak horsepower ratings. Such figures are a measurement of a vacuum's motor when it is operating freely or not under load.

"Since peak horsepower is outside the normal operating range of a vacuum cleaner, it may not be indicative of actual power differences when comparing two cleaners," warns Shop-Vac brands technical literature.

Russ Battisto at A-1 Vacuum Cleaner Company (800/657-1874 or www. a-1vacuum.com) in St. Paul, Minnesota, points out that "high-quality commercial brands may only have 1½-hp. motors, but they'll pull more vacuum, do more work, and last longer than an inexpensive brand with a much higher power rated motor."

So if horsepower isn't the ultimate measure of vacuum quality, what is?

That would be a combination of a vacuum's cubic feet per minute (CFM) and sealed pressure ratings.

MOVING VOLUMES OF AIR

CFM is a measure of the quantity or volume of air a vacuum moves when operating at a particular load. The load, in this case, refers to restriction at the vacuum's orifice as well as the type, size, and length of hose in use, as well as its filtering system. You will hear the term *velocity* occasionally used with shop vacs, but it is meaningless as a measure of machine capacity.

Sealed pressure ratings represent a tool's ability to lift water. But don't confuse this measurement for a utility vacuum's ability to suck up water off a floor. Instead, this number is commonly used by the manufacturers of industrial cleaners to describe a vacuum's suction capacity. The rating is listed in inches and describes how much a calibrated column of water is lifted against gravity using the vacuum motor.

Some manufacturers also refer to the air power performance of their machines. This rating combines the CFM and sealed pressure created at the point of operation, such as at the attachment at the end of a hose.

STAGING THE WORK

So what produces the CFM and suction a cleaner generates? That would be narrow impellers spun by a vacuum's motor. These impellers are created by sandwiching six to eight curved vanes between two thin disks that are 4 to 6 inches in diameter and spaced ¼ to ½ inch apart. When spun at high speeds, these vanes grab air from a cleaner's intake and discharge it out the exhaust port.

Most vacuum cleaners will list the number of stages they use. Each stage represents a single impeller. Smaller or cheaper vacuums typically operate with a single stage.

Larger, more expensive machines employ two and three stages. Multiple impeller stages act like jet engine turbines and compound suction for better performance.

While the suction they create can lift clods off a floor, rarely do impellers come in contact with the dirt and debris they transport. Heavier items are dropped out of the cyclonic airflow created inside a cleaner's canister much like a centrifugal air filtration system on a tractor.

What is left over (such as fine dust and light items) is stopped by a vacuum's filter before it reaches its impellers. And this makes the filtering system the second most important feature on a utility vacuum as it has a direct impact not only on a tool's performance but also on your health.

Filtering medium vary greatly among utility vacuums. At the low end of the filter spectrum are paper bags and foam coverings. Premium filters are pleated (corrugated) paper cartridges that offer more surface space to capture particles. The ultimate of all pleated filters is the HEPA version, which removes .2 to .3 micron-size particles. (A micron is .00003937 inch.)

HEPA filters have the advantage of removing 99%-plus of the particles in a vacuum's airflow. This prevents dust from attacking impellers, and jeopardizing their performance and lift. More importantly, HEPA filters remove those particles from the air and prevent them from entering your lungs.

STICK WITH QUALITY FOR BETTER DURABILITY

Certainly there are other features to consider when buying a utility vacuum: canister size, variety of attachments, and hose diameter and type. Battisto at A-1 Vacuum recommends when searching for a cleaner to use in a farm shop that you consider the premium brands.

"Many inexpensive vacuums are built to be disposable," he says. "They use small powerful motors with fewer windings that have short service life. They can self-destruct as a result of overheating or from low-quality bearings." ∎

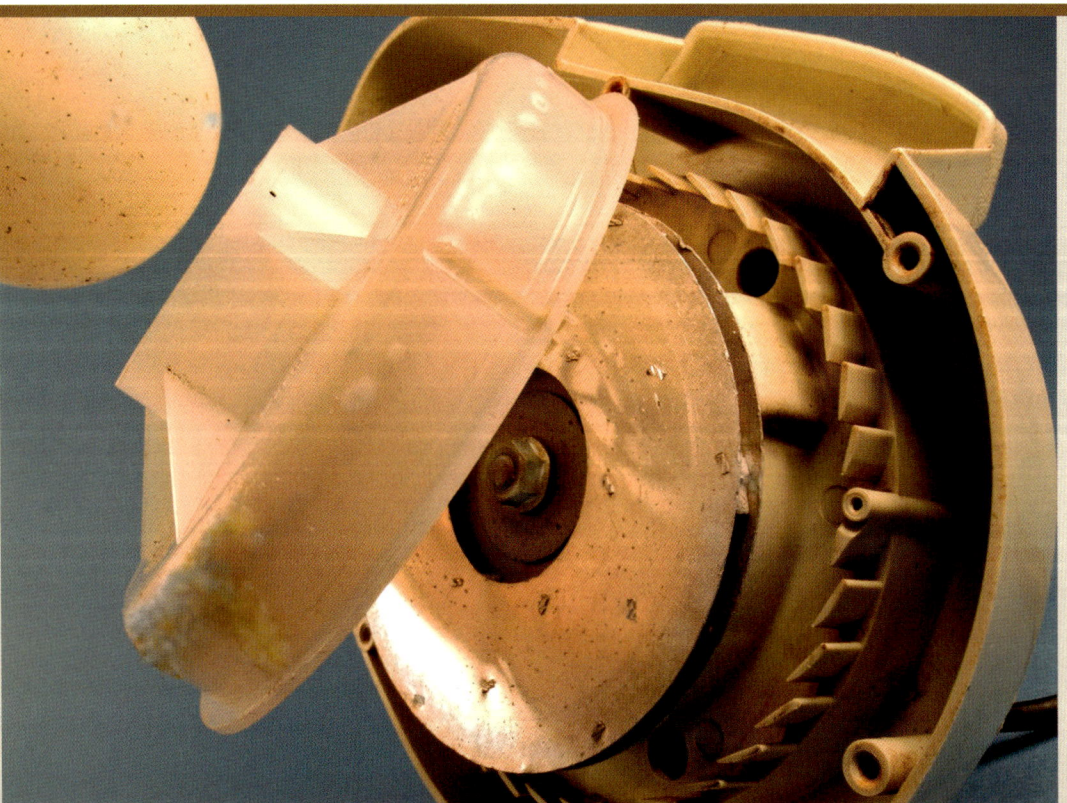

This single-stage vacuum's impeller is surrounded by a plastic housing that directs air from the intake (by the ball) from the canister. The ball works to block the intake when a vacuum is used to suction water.

Photographs: Lincoln Electric, Hobart Welders

Spool Guns

A practical guide to must-have tools for any farm shop

By John Dappert

The quality of spool gun you'll need depends on the extent of the repair. An entry-level gun is adequate if you only handle lightweight repairs on thin metal (like a ladder rung).

Anyone who's ever created an aluminum wire bird's nest in a wire welder's gun knows what a headache it can be. It's easier to unroll tangled barbed wire.

Once rare on the farm, aluminum welding is becoming more common with increasing numbers of aluminum grain boxes and livestock trailers needing repair. Farmers have taken to buying spools of aluminum electrode. But feeding aluminum wire down a welding gun's cable is like pushing a wet noodle, says Jon Ertmer of Miller Electric.

"Think of steel electrode as a hard noodle," Ertmer explains. "It has more column strength, allowing it to be more easily pushed down a cable without the wire folding over on itself when it meets resistance (like bends in the cable)."

Aluminum, being a softer metal, is like a wet noodle. Push it down a cable that has lots of bends and, well, you can get a bird's nest. The wire folds over on itself forming a clump.

Now, you can get by in welding with aluminum by keeping the gun's cable as straight as possible. "I've even seen fabricators tape broomsticks to the gun and cable to keep them as straight as possible," Ertmer says.

Or you could invest in a spool gun. As the tool's name implies, spool guns carry a supply of welding wire in small, 4-inch-diameter (or smaller) spools. That wire travels less than 12 inches (depending on model and make) from the spool to the tip of the gun, says Frank Armao of Lincoln Electric. Electricity and shielding gas (flux-core aluminum wire is a rarity) is piped down the gun's cable from an existing welder.

FLEXIBILITY IN CABLES

The beauty of spool guns is that they come with cables stretching from 12 to 50 feet and longer. That makes them ideal for repair work inside the hopper bottom of a semi-trailer truck's grain trailer. And since most spool guns work with either

Spool guns specifically configured for a welder can be readily plugged into the power and gas source for easy use (the welder's existing grounding work cable is not replaced). Other spool guns require an adapter kit and even a control box for use with other brands of wire welders. Before buying a gun, be sure to check its compatibility with your welder.

aluminum or steel wire, the tool comes in equally handy for repairing grain legs, grain tanks on combines, any work that's a long way from your welder.

What makes a spool gun an almost essential tool in any welder's arsenal is that it is "fairly simple to use and provides the benefit of being relatively inexpensive," Armao says.

ADVANCED MODELS OF GUNS ARE AVAILABLE

The spool gun market has an array of advanced, feature-rich models. A good example is Lincoln's line. The Magnum 100SG is an aluminum-only gun with few features. But for $684, you can buy the Magnum SG, which provides 25 feet of cable and an integrated wire-feed speed control.

As you would guess, the cost of such feature-rich guns escalates. For example, Miller Electric's Spoolmatic 30A lists for $1,200 but delivers a whopping 200-amp output with a 100% duty cycle, which means it can weld just about continuously.

But be warned. Not every spool gun is compatible with every make, let alone model, of welder. Many are

configured to only work with a specific brand and model or require an adapter kit or control box to make them compatible.

For example, HTP's RSG200 only works with HTP welders. But you can opt up to HTP's RSG250, which includes a control box that can be installed on virtually any brand of MIG welder.

"The gun needs to be able to communicate with the power source to operate," Ertmer explains.

Before purchasing, first check with your welder's manufacturer to determine if a gun will work with the machine you own.

FEATURE DESCRIPTIONS

As mentioned before, how a gun is equipped is what separates the men from the boys. The following is a brief description of some of the more notable feature differences.

■ **Power output.** This is listed as amperages available at a percentage of duty cycle. Duty cycle describes the percentage the gun can operate at during a 10-minute interval without overheating. Thus, a gun with a rated output of 250 amps at a 60% duty cycles means that it can oper-

ate at that maximum amperage level for six minutes straight. If all you're using a spool gun for is occasional, lightweight repairs, duty cycle is not a crucial feature. For serious fabrication, a lot of amps and long cycles are crucial.

■ **Feed rate control.** This adjustment, positioned on the spool gun, regulates how fast the wire is fed through the gun and into your weld. Rates, stated in inches per minute (IPM), vary greatly by different gun models. For example, the rate on Hobart's DP-3035 will adjust from 115 to 715 IPM, whereas their DP-3534 turns out up to a whopping 1,200 IPM.

■ **Wire capacity.** Described in inch diameters, this rating lists not only the range but also the content of the wire. For example, Lincoln's Magnum 100SG can only accommodate aluminum electrode .030" to .035" in diameter. But their 250LX feeds .023" to ³⁄₆₄" wire that is aluminum, steel, or stainless steel.

■ **Drive roll tension.** Allows you to vary the amount of resistance pressure on the roll of wire. Tension adjustment becomes crucial if you're switching from 4000 series to 5000 series aluminum wire for melding thicker material. "Being thinner, 4000 series requires half the tension of 5000," Ertmer says.

■ **Cable length.** The length of feed cables varies greatly by model. If your spot welding cracks on the outside of a livestock trailer, then a 12-foot cable might be adequate. But for repairing the inside of a combine's grain tank, consider a gun with a longer cable.

■ **Recessed, or notched, contact tip.** The difference between standard and recessed tips is that the recessed nozzle has a notch at its tip. The advantage here is that it's much easier to remove electrodes that have burned back and fused to the end of the nozzle. ■

Tips On TIG Supplies

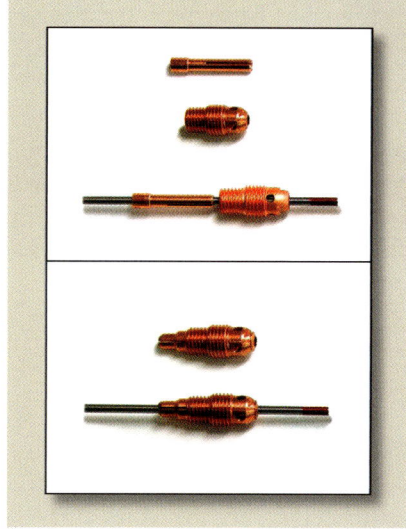

Choosing between a standard two-piece system (top) or a single-piece system (bottom) is one of the factors in determining which collets and collet bodies are best when you tackle a TIG welding job.

Back up a TIG welder with the right consumables for the best welds

By Dave Mowitz, Machinery Editor

Stick welding can spoil a person. Armed with a sleeve of electrodes, you can tackle a wide variety of basic welding chores. But increasing numbers of farmers are turning to TIG (tungsten inert gas) welding to mend an increasing diversity of metals, such as aluminum or stainless steel, or to tackle more complicated welding tasks, such as paper-thin sheet metal.

TIG welders, however, employ a smorgasbord of consumables – those bits and pieces that wear out with welding. Items such as back caps, collets and collet bodies, gas lenses, and nozzles comprise a small portion of the total cost of TIG welding, says Mike Sammons of Weldcraft. Yet, they have a huge impact on the quality of welds.

BACK CAP DIFFERENCES

Chief among TIG consumables are back caps, which apply pressure to the back end of the collet to force it against the collet body. This pressure holds the tungsten in place and seals the torch head from the atmosphere, Sammons explains. Back caps twist into the back of the torch to help create a vise that prevents the tungsten from slipping.

Back caps are made of a phenolic compound, each of which vary in temperature resistance. "You need to match the type of back cap to your application," Sammons says. "For example, using a back cap made of a low-temperature phenolic compound works for general applications. But on demanding or high-duty cycle applications, low-temperature back caps can shrink, crack, or split. So you need a back cap with a high thermal resistance for these applications to avoid weld discontinuities that can result from shielding gas leaks."

There are three types of back caps: short (button), medium, and long. A short back cap is the smallest. Its size allows for welding in restricted areas, but it takes nonstandard tungsten 2 inches or less in length. You may have to fabricate this size tungsten from a longer piece, since shorter tungsten is harder to find, Sammons warns.

If joint access is not a factor, use a medium or long back cap. A medium back cap generally accommodates tungsten up to 3 inches long. Long back caps, which are most commonly used, work with standard tungsten up to 7 inches long.

COLLETS ARE THE CONTACTS

Collets hold the tungsten in place when you tighten the back cap and create the electrical contact necessary for good current transfer. When choosing collets and collet bodies, you need to consider two main factors, Sammons advises. "First is price. Standard-grade copper collets and collet bodies cost less, but they also tend to be less durable," he explains. "Less expensive collet and collet bodies are also prone to failure under high-temperature applications. After extended use, they do not secure the tungsten as reliably."

Conversely, the more expensive tellurium copper collets and collet bodies have better heat resistance on higher amperage applications. These consumables better resist twisting or elongating to hold tungsten more securely after extended periods of use.

When selecting collets, you'll also need to decide between a single- or two-piece system. "Typically, manufacturers will sell collets and collet bodies separately to match a specific tungsten size," Sammons explains. "For example, you would purchase a ¹⁄₁₆-inch collet and collet body to match a ¹⁄₁₆-inch tungsten."

There are also single-piece systems that combine the collet and collet body together. "They provide better securing force and are easier to remove when used in a demanding application," Sammons says. "They also reduce the possibility of mismatching collet and collet body sizes, allow for quicker tungsten changeover, and help simplify parts management."

GAS LENSES LOWDOWN

Generally made of a copper and brass combination with stainless

left: Many standard gas lenses have several screens and spacers.

right: Two-piece gas lenses, which are more expensive, employ more mesh configurations or have a porous filter media in lieu of multiple screen layers.

steel screens, gas lenses replace the collet body to increase shielding gas coverage and reduce turbulence.

"The least expensive gas lenses typically have fewer screens and coarser mesh configurations, which are less durable and can negatively affect gas flow," Sammons says. "Higher quality lenses often require several layers of screens. The most durable and expensive gas lenses incorporate a filter media in lieu of multiple screen layers. This type of filter media gives the best performance."

Application and performance dictate lens choice, Sammons adds. When welding material that tends to react to atmospheric contaminants, larger gas lenses provide improved gas coverage. "On complex joints, larger lenses also allow greater tungsten stickout to gain visibility of the weld puddle or to increase the access to the joint," he says.

For example, you could use a standard or a large gas lens on a 17 series air-cooled torch or on an 18 series water-cooled torch. For a 9 series air-cooled torch or a 20 series water-cooled torch, a standard or large lens provides good gas coverage.

DELIVERING THE GAS

Nozzles (also called cups) provide a given amount of shielding gas coverage to the weld, according to their size. For example, a smaller nozzle provides less gas coverage than a larger one. Nozzles also vary in length, price, and performance, Sammons says.

The most cost effective are 90% or 95% alumina oxide nozzles, which work adequately for lower amperage applications. These nozzles, however, don't resist thermal shock on higher amperage applications well and tend to deteriorate or crack and fall off.

Lava nozzles cost more than alumina oxide but are more resistant to cracking, Sammons points out. These nozzles work well on medium-amperage applications but tend to have varying wall thicknesses around the inside diameter, a factor that may lead to unequal gas coverage.

THE CADILLAC OF NOZZLES

Silicon nitrate nozzles are the most expensive nozzles and also the highest performing. These nozzles resist heat and cracking on higher amperage and duty-cycle applications and last longer than lava or alumina oxide nozzles. "For precision TIG welding, silicon nitrate offers the consistency and durability needed to achieve quality welds and avoid rework," Sammons says. "In fact, the higher initial purchase price for silicon nitrate nozzles may be worthwhile to avoid the ongoing cost of replacing inexpensive, lesser quality nozzles."

All in all, when shopping for TIG consumables, Sammons suggests cheaper items when just adequate performance is needed. "If you need higher performance welds, it's worth it to purchase more expensive, long-lasting consumables," he says. "It minimizes the chance of consumable failures that could increase the cost for rework with weld problems." ■

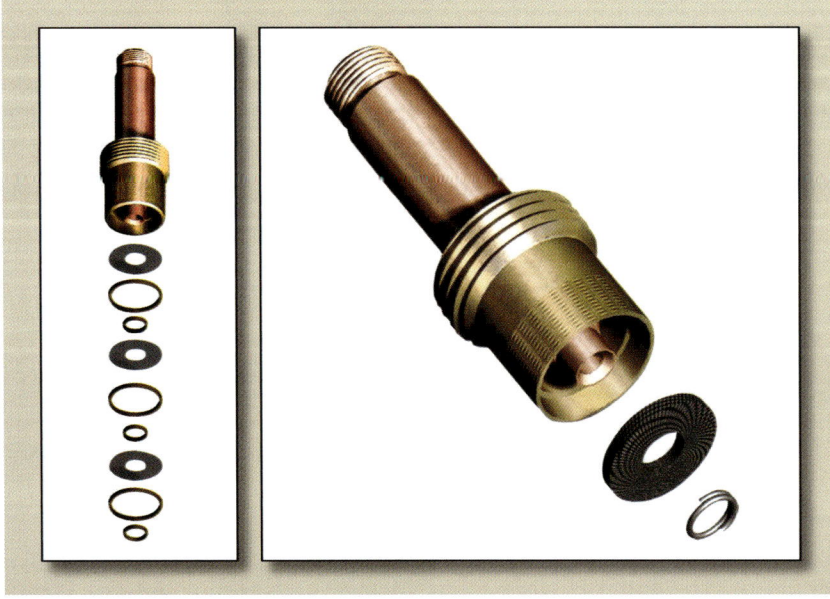

Welding Equipment

Photographs: Manufacturers   Chart: Hobart Welders

Before you begin to repair a piece of equipment, know what type of metal you're working with. By choosing the proper tool, you will make a better weld.

With the right welder and eye protection, you're ready to make repairs in the shop or in the field

By Laurie Potter, New Products Editor

When you've got equipment in need of repair, how do you decide what type of welder will do the job? For most farmers, the tool they choose comes down to four things: budget, simplicity, convenience, and portability.

Narrow the choices down by knowing how you'll use the machine.

■ Will you be welding in the shop mainly, or will you be traveling to the field for a quick repair?

■ What type and thickness of metal will you be welding?

SELECTING THE TOOL FOR THE TASK

Stick welders are popular because they work well in windy conditions and on rusty or oily materials. They're also the least expensive.

MIG/flux-cored welders use a spool of wire continuously fed through a gun. These welders eliminate the need to replace electrodes when they burn down, and they work better on sheet metal. The process is also easier to learn.

TIG welders perform best on precision work, like engines, stainless steel pipe on dairy farms, or very thin metal. They use the slowest process and require the most skill.

Some welders can do more than one task, such as a welder-generator combination or a MIG/TIG/stick welder in one unit.

Welder prices vary considerably, running from a few hundred dollars to several thousand dollars depending on options, voltage, etc.

Before you purchase a welder, do some research to find the proper tool for the welding you'll be tackling.

HELMET SELECTION

Before you ever fire up the welder, you need to have the proper eye protection, of course. Take a look at these two types of helmets:

■ **Standard** (or passive) helmets are made of molded plastics. The viewing lens, or filter, is a special piece of dark-tinted glass. These welding helmets flip up and down and give the

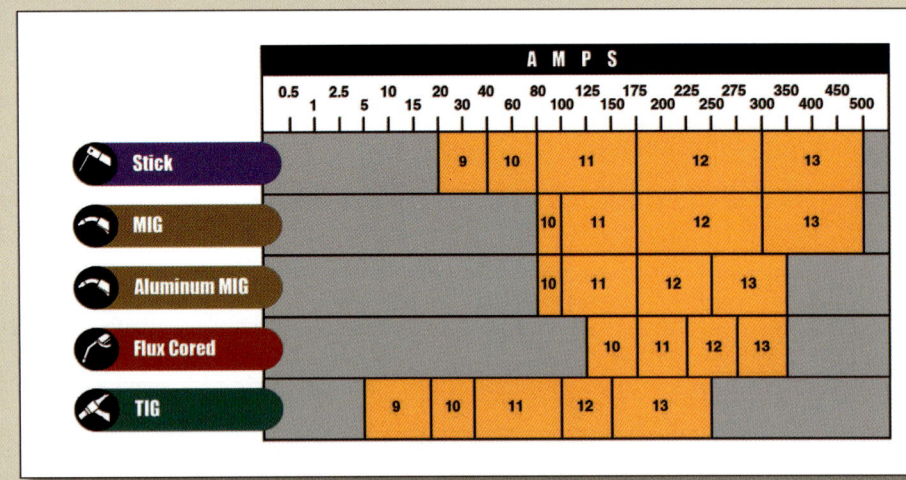

AMPS																			
	0.5		2.5		10		20		40		80		125	175		225	275		350	450
		1		5		15		30		60		100		150		200		250	300	400	500
Stick								9		10		11			12			13	
MIG										10		11			12			13	
Aluminum MIG										10		11		12		13			
Flux Cored												10	11		12	13			
TIG						9		10		11		12		13					

This helmet shade rating chart identifies the amount of shade needed for a specific welding application. The rating (9 to 13) is based on the amps the unit discharges and tells you the recommended shading needed to protect the operator's eyes.

wearer basic protection. They range in price from $20 to $30. Because these helmets have to be lowered and raised, they can make welding in tight places difficult.

■ **Auto-darkening** helmets have an electronic filter lens and often come equipped with adjustable features such as a fixed or variable shade.

If most of your welding involves one type of metal, the fixed shade is all you'll need. But if you work on various types of metal where the welding amperage varies from 40 to more than 200 amps, a variable-shade helmet may be an option to consider.

Be sure the helmet you purchase has an ANSI (American National Standards Institute) rating of Z87.1 - 2003 or Z87+. ■

GUIDE TO SELECTING A WELDING HELMET LENS

Many welders mistakenly think that the lens shade number on welding helmets corresponds to the amount of protection provided to eyes. "The idea is that the higher the number, the better the protection," says Jim Harris of Lincoln Electric Company.

All helmets that comply with industry standards filter out a vast majority of the harmful ultraviolet (UV) and infrared radiation (IR) emissions to protect the eyes. The choice in lens number is simply a matter of comfort for the user, Harris explains. "Shade number denotes the level of shading provided by that particular lens and should be used as a guide to select a lens that is comfortable and provides good visibility."

MUST BE ABLE TO SEE THE WELD

Always select a shade that allows you to see the weld puddle clearly, Harris recommends. When in doubt, refer to manufacturers' recommendations for lens shade numbers that fit various applications.

When making a lens selection, you must understand the meaning of arc flash and what the different types of emissions are that radiate from the welding arc. Arc flash is simply the unexpected exposure of the eyes to the welding arc. The welding arc emits several forms of light, including UV, IR, and high-intensity visible light. Both UV and IR rays can cause permanent eye damage (such as retinal burns).

While high-intensity visible light may not cause permanent eye damage, it may leave the operator with temporary discomfort similar to being exposed to the flash of a camera bulb.

LOOK FOR ANSI RATING

All helmets (when in the down position) that comply with the current American National Standards Institute standard (ANSI Z87.1) protect operators from the harmful UV and IR damaging elements of the arc. When selecting a helmet, look for a helmet that is certified by the manufacturer to comply with this standard, Harris explains. Also, look for this labeling on autodarkening cartridges to confirm that the helmet does comply with this standard. ■

Master The Sandblaster With True Grit

Many people can misinterpret the intentions of a man in a mask and a hood

By Roger Welsch

I don't know how to explain my pleasure at cleaning parts. Sitting at my workbench, picking away at carbon bits, dirt, and grease is such a quiet, harmless sort of activity. And it's best done on a rainy or snowy day when I can occasionally look out my window and just be glad to be alive and in my warm, comfortable shop.

I'm not quite as tickled with corroded manifolds or irretrievably caked oil pans. They're too big to be any fun, and the dirt won't come off, even with a power tool. That dirt is permanent. Or, at least, it seems to be permanent. Or, it used to seem permanent.

But it isn't. A sandblaster will take it off. Like magic. And not just take off the dirt. But leave it as shiny as it was when it was brand-new.

Now, after a year or so of working with my very low-tech, inexpensive sandblasting outfit, I have learned enough about the process that I figure I can pass along some basic information to you in the unlikely event that you know even less than I do. (In which case, you should be embarrassed, if not downright ashamed.)

For example, I've found it a bad idea to eat a peanut butter sandwich while sandblasting unless your family doctor has suggested that you could use more fiber in your diet. A lot more fiber.

PROTECTIVE EQUIPMENT

Of course, a large part of sandblasting is good, protective equipment. When sandblasting, it's a good idea (a *very* good idea) to wear gloves, although there is the built-in advantage of cleaning your fingernails at the same time you're removing rust from an old PTO shaft. Be sure to wear good, tight goggles – or better yet, a full face mask. Otherwise, you'll be awake all night listening to your eyelids scrape across your eyeballs.

A couple of words of caution from bitter experience: While wearing a sandblasting hood, do not attempt to make a withdrawal from your savings account, even in the local bank where everyone knows you by your first name and should know by now what an idiot you are. And, don't walk past three nervous black Labrador retrievers with the hood on.

I know that showing up in elbow-length gloves, an apron, and a full-head hood just might reinforce the impression in town or with your wife that you really are working and that you really are a mechanic. But believe me, you'll be surprised at how many people can misinterpret the intentions of a man in a mask and hood.

I once did some sandblasting while wearing shorts. Then I walked to town for a beer. This was not a good idea. I'd prefer not to go into the details, and believe me, you would prefer that, too. But just let me say that I walked funny for a long time thereafter. And never sandblasted again in my shorts.

No matter what you use for a grit, or what air pressure you use to blow it onto the part you are cleaning, let me offer one more bit of advice, and this comes from my heart: Whatever else you do, don't hold the part you are sandblasting in your lap while you're doing the work. Especially if you are wearing shorts. (As I said, I did this. Want to see my scars?)

OTHER PRECAUTIONS

If you use sand directly from your child's sandbox, or for that matter from your own, be sure to sift out all the cat poop. And don't tell me there isn't any cat poop in there because if there is a cat within 50 miles of your house (believe me, there is a cat within 50 miles your house; in fact, I'm betting there is a cat watching you from the bushes at this very moment, just waiting to poop in your sandbox) there IS cat poop in your sandbox. ∎

Photograph: Dave Mowitz

The Joys Of Owning A MIG Welder

Rog's family "benefits" from his newfound skills

By Roger Welsch

If you've ever done any welding, you know it really is fun melting metal-sticking things together, even if it can hurt really bad now and then. I recently acquired a MIG welding outfit (MIG = Multi-Injury Gizmo) and got to playing around with it. It makes welding child's play.

And you will understand that all the more if you just sit down some time with a child, give the little tyke two or three warm chocolate bars and a jar of jelly, and watch the action for a couple of hours.

In fact, I have made myself quite a hero in my home and something of a landmark in the world of welding by solving even the most unlikely problems with my welding torch and new MIG unit by making things for my loved ones!

This first started a couple years ago when my daughter, Antonia, was moping around because all her friends had in-line skates, but she didn't. And I, noting that even the discount price for these toys was nearly $100, decided that I would just go out to the shop and weld something up!

DREAMS CAN COME TRUE

I was thoroughly pleased with my solution: a pair of in-lines that cost almost nothing since I used scrap metal. (The visit to the emergency ward doesn't really count because that kind of burn could have happened eventually no matter what I was working on.) And, as I told Antonia, given a little time, those wheels will wear out round.

As it turned out, just about the time I brought the first skate into the house, Antonia decided she really wasn't interested at all in in-line skating. So I wound up sending the sample to my other daughter, Joyce, in Seattle, Washington, who opted not to ask me to complete the second, since UPS charges something like $20 for a package weighing 30 pounds.

It was perhaps a year later that my skills were next called to the front line, and this time in the name of romance. It was a couple days before St. Valentine's Day, and as yet I had found nothing for my wife, Linda.

I thought about my options. Flowers wilt and are thrown out in a day or so and next year you just have to buy more. I have found that while chocolates are always welcome, I have to listen to months of complaints about what they do to hip padding.

So, I was looking for something more permanent than flowers, less fattening than chocolates, but dripping with romance. That's when it occurred to me to go out to the shop and weld something up!

The result was a 40-pound, ¼-inch steel heart, complete with initials on the top, opening up to reveal . . . ta-da! It's plum packed full of jerky! And not the homemade kind, complete with hair and teeth. No, this was store-bought jerky. Your high-class mesquite kind, barbecue variety, and, for the exotic taste, teriyaki!

As you can imagine, Linda was so stunned, she cried like a baby and couldn't find the words to express her appreciation for almost a full week. She loves it so much, she asked me to keep it out in the shop where it will remind me of how much I love her. I think that's why she said I should keep it out in the shop anyway.

LOVE AND GRATITUDE

Who knows what's next? All I know is that my shop's welding equipment is no longer just for fun and profit. It's a surefire device for bringing me – the expert welder – the gratitude and love of my entire family.

And I'm sure Linda, Joyce, and Antonia would be the first to tell you that if they were in here . . . but I can see they're out there in the yard. Looks like they are trying to build a fire up against the west side of the shop, right in that pile of brush and cardboard. Hmmmmm. Wonder what that's all about. ■

Must-Have Tools

The next big shop items
By Roger Welsch

I n a recent column, I offered the tractor industry some ideas that will make them a fortune while earning me not a dime. Well, now I'm going to do the same thing for the world's tool manufacturers. It's my contribution to the wheels of American industry.

Anyone who wants to get rich is welcome to help him or herself to one or several of the following items. I would appreciate a couple freebies and maybe $1 or $2 million as the profits roll in.

MY LIST OF IDEAS

For example, why hasn't someone come up with a Kevlar shop apron? Full body armor for the shop would be even better – if you do your shop work like I do. Or, as I like to think of it, full-contact mechanicking.

An item like that would be valuable if someone takes up my idea for Stuck Nut Blasting Powder. You already know that patience, penetrating solvents, and gentle persuasion only go so far. Then it's time for a generous dose of Mechanic's Blaspheme Lotion and a pound or two of Stuck Nut Blasting Powder.

I would like to see a screwdriver developed that exudes just a small dose of disinfectant and coagulant every time you put even a little extra pressure on it. Same with crescent wrenches and pry bars. This would save that annoying extra step of going to the house to wash out and sterilize the inevitable lacerations between jobs.

ADD-ONS AND OPTIONS

A floor creeper with a hydraulic lift in it would be nice for us geezers who find the hardest thing about getting down to work under a tractor

Feel free to put any of Roger's great tool inventions on the market. He'll be waiting for the royalty checks at his home office.

is getting out and up from the job. A heated pad would also be nice. And maybe one of those Magic Fingers massagers.

How about a Smart Hammer with a camera so you could see every single time you swing the tool precisely where, on your thumb, it is going to land?

I do have a torque wrench. But a lot more useful (considering my level of expertise) would be a thoroughly torqued-off throwing wrench, a nicely formed tool made especially for throwing at the wall or out the door at those moments when everything else has failed.

Like when the bolt has twisted off or the pin has dropped down into the transmission for the one hundred seventy-third time.

THE BRAINSTORM JUST KEEPS STORMING

I have many other ideas, believe me. I imagine, for example, a Wife Come-Along. This is for convincing She Who Must Be Obeyed that nothing could be more romantic than taking in a tractor show together or an auction on that oh-so-important anniversary.

They already have devices for balancing wheels, and harmonic balancers for balancing . . . well, whatever a harmonic balancer balances. How about a handy-dandy checkbook balancer designed to avoid the nasty repurcussions from overdrafts that come along when you've been shopping on eBay for parts and accidentally wind up buying two more tractors?

OK, I'm weary now of making the rest of you rich. Time to hit the hammock and wait for those royalty checks to start coming in. ■

The Ultimate Shop First Aid Kit

Relief for self-inflicted wounds

By Roger Welsch

Not long ago, in a cardiac ward somewhere in America's heartland, a nurse was making her thirty-seventh futile try to find a vein in my arm into which she intended to place a ¾-inch plumbing faucet. Lovely Linda watched the process for a while before laconically and accurately noting, "Give him a butter knife, and in three minutes you'll have all the blood you want."

OK, so I'm a trifle accident-prone. And yes, I'm a bleeder. Also a bruiser. And a breaker. And a burner. So it shouldn't be a surprise that one of the first things I did when I built my own shop was to buy a large, industrial-grade first aid kit. It wasn't one full day before I realized that, industrial-grade or not, the world of medical science hasn't kept up with me when it comes to self-inflicted wounds.

I have, therefore, begun to assemble my own shop medical kit. I realize that you probably have your own specific needs when it comes to banging yourself up, and you'll need items specific to those needs. So let the list below be a start. ∎

THE ROGER WELSCH HOPELESS KLUTZ SHOP FIRST AID KIT

☐ Hernia trusses – Assorted sizes for every part of your body, including eyeball hernias from that new *Sports Illustrated* swimsuit calendar.

☐ Ice pack – Don't skimp on this one. A 50-pound potato sack should be just about right.

☐ Tetanus-booster syringes – 1 gross.

☐ Full-body gauze pads.

☐ Sanitized hospital-grade duct tape.

☐ Bulk antiseptic system – 60 gallons with electric pump should get you started.

☐ Oxygen tanks – The tank from your gas welding outfit will do, but be sure you have the hoses straight and you are sucking on the oxygen hose, not the acetylene outlet. Especially if you smoke.

☐ Bulk burn ointment – 30-gallon tub at minimum; a second as optional backup if you do any welding.

☐ Stretcher/gurney – If shop is more than 30 feet from your house, you may want a motorized version.

☐ Crutches – An assortment of sizes and grades.

☐ Surgical-grade Vise-Grip pliers – For removing splinters from various body parts as well as plucking sand, wood, bolts, or pistons from your eye.

☐ One sterile come-along – For the stuff stuck in you that the Vise-Grips won't pull out.

☐ Eye rinse – 60 gallons or a fire hydrant.

☐ 2×4s for splints – Several hundred board feet.

☐ Large mirror – Allows you to see personal damage where you sat on that newly welded side rail. Note: While I do recommend a mirror, I strongly urge you not to include a magnifying glass in your medical kit. If you're in my league, all personal damage will be big enough to see, and you won't want to see it any closer.

☐ Paint brushes – Use these to apply ointments and disinfectant. For those larger wounds, a paint roller may work better.

☐ Jack Daniels Green Label – Deadens pain from external wounds and lifts the internal spirit.

☐ Cell phone – Predial it to 911. ∎